AF411338

New Challenges in Wood and Wood-Based Materials

New Challenges in Wood and Wood-Based Materials

Editors

Ľuboš Krišťák
Roman Réh
Ivan Kubovský

MDPI • Basel • Beijing • Wuhan • Barcelona • Belgrade • Manchester • Tokyo • Cluj • Tianjin

Editors

Ľuboš Krišťák
Department of Physics,
Electrical Engineering and
Applied Mechanics
Technical University in Zvolen
Zvolen
Slovakia

Roman Réh
Department of Wood Technology
Technical University in Zvolen
Zvolen
Slovakia

Ivan Kubovský
Department of Physics,
Electrical Engineering and
Applied Mechanics
Technical University in Zvolen
Zvolen
Slovakia

Editorial Office
MDPI
St. Alban-Anlage 66
4052 Basel, Switzerland

This is a reprint of articles from the Special Issue published online in the open access journal *Polymers* (ISSN 2073-4360) (available at: www.mdpi.com/journal/polymers/special_issues/Wood_Based_Mater_2021).

For citation purposes, cite each article independently as indicated on the article page online and as indicated below:

LastName, A.A.; LastName, B.B.; LastName, C.C. Article Title. *Journal Name* **Year**, *Volume Number*, Page Range.

ISBN 978-3-0365-1792-6 (Hbk)
ISBN 978-3-0365-1791-9 (PDF)

This publication was supported by the Slovak Research and Development Agency under contract No. APVV-18-0378, APVV-19-0269 and VEGA1/0717/19.

Contents

 *polymers*

Editorial

New Challenges in Wood and Wood-Based Materials

Lubos Kristak *, Ivan Kubovský and Roman Réh

Faculty of Wood Sciences and Technology, Technical University in Zvolen, T.G. Masaryka 24, 96001 Zvolen, Slovakia; kubovsky@tuzvo.sk (I.K.); roman.reh@tuzvo.sk (R.R.)
* Correspondence: kristak@tuzvo.sk

check for updates

Citation: Kristak, L.; Kubovský, I.; Réh, R. New Challenges in Wood and Wood-Based Materials. *Polymers* **2021**, *13*, 2538. https://doi.org/10.3390/polym13152538

Received: 25 June 2021
Accepted: 26 July 2021
Published: 31 July 2021

Publisher's Note: MDPI stays neutral with regard to jurisdictional claims in published maps and institutional affiliations.

Wood and wood-based composites are key engineering materials that can be successfully designed and manufactured with predetermined exploitation properties, making them suitable for a wide range of applications and end uses. Notably, wood-based composites can be engineered to meet specific performance requirements, which makes them a sustainable solution for reducing the use of solid wood.

In the difficult times caused by the COVID-19 pandemic, we succeeded in compiling another interesting publication. It is a Special Issue of *Polymers* (ISSN 2073-4360) that belongs to the Section *"Biomacromolecules, Biobased and Biodegradable Polymers"* of MDPI. Even before the widespread outbreak of the pandemic, not knowing what was coming, the Special Issue was given the title "New Challenges in Wood and Wood-Based Materials". Then the new challenges really came. We initially referred to different challenges. We certainly did not imagine that many researchers would have difficulty in routine scientific research work as a result of the significant disruption. The mobility of people has been restricted, access to workplaces has been more limited, and the preparation and compilation of scientific papers has become more complicated. Meeting the deadlines associated with the peer review process has been much more demanding than during normal times.

This Special Issue of *Polymers* is a collection of 11 original high-quality scientific contributions on basic and applied research in the field of wood science and technology, and provides good examples of the recent challenges related to the production and application of wood and wood-based materials. Individual papers concerned with the enhancement of the performance and technological properties of wood composites, above all plywood [1–3], as well as with ignition and combustion of wood and wood composites in monitoring and evaluating these processes on state-of-the-art equipment [4–7], and monitoring chemical changes in wood and wood adhesives and composites [8–11], are included. The topic of the Special Issue has clearly resonated in the world's scientific community and the traditional response has come from strong wood research centers in Europe and Asia.

This Special Issue *"New Challenges in Wood and Wood-Based Materials"* follows up our previous Special Issue *"Application of Wood Composites"* in *Applied Sciences* [12]. In the first Special Issue, it turned out that wood and wood composite materials are engineered materials with significant physical, mechanical, or chemical properties which predetermine them for a whole range of uses, known or not yet discovered. Wood composites with their outstanding properties have replaced some conventional materials in various fields of applications. However, there is now a new and bolder goal: challenges. What are challenges? Something that by its nature or character serves as a call to make a special effort, a demand to explain or justify, or a difficulty in an undertaking that is stimulating to one engaged in it. Is this possible within the basic or applied research dealing with wood science and technology? The second Special Issue says quite clearly that we can also move within such boundaries of areas.

The scientific goal of this publication is to provide the reader with new information on recent practices in plywood research. Yes, plywood, as plywood is definitely back. Towards the end of the 20th century and also the beginning of the 21st century, there was a clear decline in plywood research. However, its unexpected increase in production and the expansion of its application also caused an increase in research activities. This is also reflected in this Special Issue. The study [1] investigated the effect of phenol-formaldehyde resin treatment on the weathering stability and biological durability of birch plywood. Silver birch veneers were vacuum-pressure-impregnated with four different phenol-formaldehyde resins with average molecular weights. The aging properties of phenol-formaldehyde resin-modified birch plywood were analyzed using artificial weathering with ultraviolet light, water spray, and weathering under outdoor conditions. The same combinations of phenol-formaldehyde-treated plywood specimens were then tested in soil-bed tests to determine their resistance against soft-rot wood decay. The weathering stability of birch plywood treated with phenol-formaldehyde resins has been proven. Results from unsterile soil-bed tests showed improvements in resistance to soft-rot wood decay compared to untreated plywood and solid wood.

A remarkable challenge is the utilization of large-scale wood in existing additive manufacturing techniques [2]. Wood-based materials in current additive manufacturing feedstocks are primarily restricted to the micron scale. This study proposes an additive manufacturing method—laser-cut veneer lamination—for wood-based product fabrication. Inspired by laminated manufacturing and common plywood technology, laser-cut veneer lamination bonds wood veneers in a layer-upon-layer manner. As demonstrated by printed samples, laser-cut veneer lamination was able to retain the advantageous qualities of additive manufacturing, specifically, the ability to manufacture products with complex geometries which would otherwise be impossible using subtractive manufacturing techniques. Furthermore, laser-cut veneer lamination product structures designed through adjusting internal voids and wood texture directionality could serve as material templates or matrices for functional wood-based materials. Numerical analyses established relations between the processing resolution of laser-cut veneer lamination and proportional veneer thickness (layer height).

The third work on plywood research deals with three dimensionally molded plywood formed parts [3]. They were prepared in two different geometries using cut-outs and relief cuts in the areas of the highest deformation. The effect of flax fiber reinforcement on the occurrence and position of cracks, delamination, maximum load capacity, and on the modulus of elasticity was studied. The results show that designs with cut-outs are to be preferred when molding complex geometries and that flax fiber reinforcement is a promising way of increasing the load capacity and stiffness of plywood formed parts.

Wood and wood-based materials can be subject to combustion, and therefore research into reducing their flammability requires long-term attention. This is also evident in this Special Issue. The study [4] focuses on the energy potential and combustion process of torrefied wood. Samples were prepared through the torrefaction of five types of wood: ash, beech, oak, pine, and spruce. They were heated under a nitrogen atmosphere. The samples enabled the investigation of torrefied wood combustion in a compact form. The effect of the external heat flux on the combustion of the samples was measured using a cone calorimeter. The observed parameters included initiation times, heat release rate, and combustion efficiency. The results show that increasing the external heat flux decreases the evenness of combustion of torrefied wood. At the same time, it increases the combustion efficiency. The results are useful both for the energy production field and for fire safety risk assessments of stored torrefied wood.

An experimental study of straw-based eco-panel using a small ignition initiator [5] points to another fundamental problem. It is well known that straw, a natural cellulose-based material, has become part of building elements and eco-panels, and compressed straw in a cardboard casing is used as building insulation because of its excellent insulating properties. If suitably fire-treated (insulation and covering), straw panels' fire resistance may be increased. This study deals with monitoring the behavior of eco-panels exposed to a small ignition initiator (flame). The samples consisted of compressed straw boards coated with cardboard. Samples were exposed to a flame for 5 and 10 min. The influence of the selected factors (size of the board, orientation of flame to the sample) was compared on the basis of experimentally obtained data of mass loss. The results obtained do not show a statistically significant influence of the position of the sample and the initiating source (flame). The results presented in the paper confirm the justifiability of fire tests. As the results of the experiments prove, the position of a small burner igniting such material is also important. This weakness of the material can also be eliminated by design solutions in their construction. The experiment on larger samples also confirmed the justifiability of fire tests along with the need for flame retardancy of such a material for its safe application in construction.

If we talk about the protection of wood-based materials against fire, this area clearly includes oriented strand boards (OSBs). The experimental study of OSB ignition by radiant heat fluxes [6] investigated the ability of materials to ignite when heated at elevated temperatures. It depends on many factors such as the thermal properties of materials, the ignition temperature, critical heat flux, and the environment. OSBs without any surface treatment were used as experimental samples. The samples were gradually exposed to a heat flux. The ignition times are similar for all OSB thicknesses. The influence of the selected factors (thickness and distance from the heat source) was analyzed based on the experimentally obtained data of ignition time and weight loss. The results show a statistically significant effect of OSB thickness on ignition time.

The fire-technical properties of common woods such as spruce, oak, beech, etc. used in different constructions and buildings have been given much attention, while less attention has been paid to tropical woods as they represent more complex systems [13–15]. Selected tropical wood species (cumaru, garapa, ipe, kempas, merbau) were tested from the point of view of non-isothermal thermogravimetry [7]. For these tropical woods, a relationship was established between non-isothermal thermogravimetry runs and the wood weight loss under flame during cone calorimetry flammability testing. A correlation was found for the rate constants for decomposition of wood found from thermogravimetry and the total time of sample burning related to the initial mass. Non-isothermal thermogravimetry runs were assumed to be composed of three theoretical runs, such as decomposition of the wood into volatiles, oxidation of carbon residue, and the formation of ash. A fitting equation of three processes was proposed and the resulting theoretical lines match experimental lines.

Other selected tropical woods (meranti, padauk) and merbau were tested from the point of the view of their changes in wood lignin during the ThermoWood process [8]. Thermal modification is an environmentally friendly process in which technological properties of wood are modified using thermal energy without adding chemicals, the result of which is a value-added product. Wood samples of three tropical wood species were thermally treated according to the ThermoWood process at various temperatures (160, 180, 210 °C) and changes in isolated lignin were evaluated by nitrobenzene oxidation (NBO), Fourier transform infrared spectroscopy (FTIR), and size exclusion chromatography (SEC). New data on the lignins of the investigated wood species were obtained, e.g., syringyl to guaiacyl ratio values. Higher temperatures cause a decrease in methoxyls and an increase in C=O groups. Simultaneous degradation and condensation reactions in lignin occur during thermal treatment, the latter prevailing at higher temperatures.

The chemical composition and morphological properties of Norway spruce wood and bark were evaluated in [9]. The extractive, cellulose, hemicellulose, and lignin contents were determined by wet chemistry methods. The dimensional characteristics of the fibers (length and width) were measured by Fiber Tester. The results of the chemical analysis of wood and bark show the differences between the trunk and top part, as well as in the different heights of the trunk and in the cross section of the trunk. The biggest changes were noticed between trunk bark and top bark. Fiber length and width depend on the part of the tree, while the average of these properties is larger depending on the height. Both wood and bark from the trunk contain a higher content of fine fibers and a lower content of longer fibers compared to the top. During storage, a decrease in extractives occurred, mainly in bark. Wood from the trunk retained very good durability in terms of chemical composition during the storage. In view of the morphological characteristics, a decrease in both the average fiber length and width in wood and bark occurred.

The growing need for sustainable products and the stringent legislative requirements related to the hazardous formaldehyde emissions from wood-based panels have boosted scientific and industrial interest in the production of eco-friendly, wood-based panels and optimal utilization of the available lignocellulosic materials [16–21]. The potential of the production of eco-friendly, formaldehyde-free, high-density fiberboard (HDF) panels from hardwood fibers bonded with urea-formaldehyde (UF) resin and a novel ammonium lignosulfonate (ALS) is investigated in the paper [10]. HDF panels were fabricated in the laboratory by applying a very low UF gluing factor (3%) and ALS content varying from 6% to 10% (based on the dry fibers). The physical and mechanical properties of the fiberboard, such as water absorption, thickness swelling, modulus of elasticity, bending strength, and internal bond strength, as well as formaldehyde content, were determined in accordance with the corresponding EU standards. The HDF panels exhibited very satisfactory physical and mechanical properties, fully complying with the standard requirements of HDF for use in load-bearing applications in humid conditions. Markedly, the formaldehyde content of the laboratory-fabricated panels was extremely low, ranging from 0.7–1.0 mg/100 g, i.e., meeting the most stringent requirements of the super E0 emission grade ($\leq$1.5 mg/100 g), which allowed their classification as eco-friendly, low-emission, wood-based composites.

Last, but not least, the impact of wood waste on the mechanical and biological properties of silicone-based composites was investigated using wood waste from oak, hornbeam, beech, and spruce trees [11]. The density, abrasion resistance, resilience, hardness, and static tensile properties of the obtained wood–plastic composites were tested. The results revealed slight changes in the density, increased abrasion resistance, decreased resilience, increased hardness, and decreased strain at break and stress at break compared with untreated silicone. The samples also showed no cytotoxicity to normal human dermal fibroblasts. The possibility of using the prepared composites as materials to create structures on the seabed was also investigated by placing samples in a marine aquarium for one week and then observing sea algal growth.

We would like to thank to our section managing editor of the section *"Polymer Chemistry"*, Chris Chen, for all his assistance and ongoing support throughout the publishing process.

As the topic *"New Challenges in Wood and Wood-Based Materials"* is still relevant, i.e., there are emerging new challenges in wood and wood-based materials, it is understandable that MDPI has already opened submissions to a new Special Issue *"New Challenges in Wood and Wood-Based Materials II"* within the journal *Polymers* with the possibility of publishing work on a new wide range of wood and wood composite material challenges. We will be grateful for your further excellent scientific papers.

Acknowledgments: This publication was supported by the Slovak Research and Development Agency under contract No. APVV-18-0378, APVV-19-0269 and VEGA1/0717/19.

Conflicts of Interest: The authors declare no conflict of interest.

References

1. Grinins, J.; Biziks, V.; Marais, B.N.; Rizikovs, J.; Militz, H. Weathering Stability and Durability of Birch Plywood Modified with Different Molecular Weight Phenol-Formaldehyde Oligomers. *Polymers* **2021**, *13*, 175. [CrossRef]
2. Tao, Y.; Yin, Q.; Li, P. An Additive Manufacturing Method Using Large-Scale Wood Inspired by Laminated Object Manufacturing and Plywood Technology. *Polymers* **2021**, *13*, 144. [CrossRef]
3. Jorda, J.; Kain, G.; Barbu, M.-C.; Haupt, M.; Krišťák, Ľ. Investigation of 3D-Moldability of Flax Fiber Reinforced Beech Plywood. *Polymers* **2020**, *12*, 2852. [CrossRef] [PubMed]
4. Rantuch, P.; Martinka, J.; Ház, A. The Evaluation of Torrefied Wood Using a Cone Calorimeter. *Polymers* **2021**, *13*, 1748. [CrossRef]
5. Makovicka Osvaldova, L.; Markova, I.; Jochim, S.; Bares, J. Experimental Study of Straw-Based Eco-Panel Using a Small Ignition Initiator. *Polymers* **2021**, *13*, 1344. [CrossRef] [PubMed]
6. Tureková, I.; Marková, I.; Ivanovičová, M.; Harangózo, J. Experimental Study of Oriented Strand Board Ignition by Radiant Heat Fluxes. *Polymers* **2021**, *13*, 709. [CrossRef]

7.   Makovicka Osvaldova, L.; Janigova, I.; Rychly, J. Non-Isothermal Thermogravimetry of Selected Tropical Woods and Their Degradation under Fire Using Cone Calorimetry. *Polymers* **2021**, *13*, 708. [CrossRef]
8.   Kačíková, D.; Kubovský, I.; Gaff, M.; Kačík, F. Changes of Meranti, Padauk, and Merbau Wood Lignin during the ThermoWood Process. *Polymers* **2021**, *13*, 993. [CrossRef]
9.   Čabalová, I.; Bélik, M.; Kučerová, V.; Jurczyková, T. Chemical and Morphological Composition of Norway Spruce Wood (Picea abies, L.) in the Dependence of Its Storage. *Polymers* **2021**, *13*, 1619. [CrossRef] [PubMed]
10. Antov, P.; Savov, V.; Krišťák, Ľ.; Réh, R.; Mantanis, G.I. Eco-Friendly, High-Density Fiberboards Bonded with Urea-Formaldehyde and Ammonium Lignosulfonate. *Polymers* **2021**, *13*, 220. [CrossRef]
11. Mrówka, M.; Szymiczek, M.; Skonieczna, M. The Impact of Wood Waste on the Properties of Silicone-Based Composites. *Polymers* **2021**, *13*, 7. [CrossRef]
12. Krišťák, Ľ.; Réh, R. Application of Wood Composites. *Appl. Sci.* **2021**, *11*, 3479. [CrossRef]
13. Makovicka Osvaldova, L.; Osvald, A. Flame Retardation of Wood. *Adv Mat Res* **2013**, *690–693*, 1331–1334.
14. Čekovská, H.; Gaff, M.; Osvaldová, L.; Kačík, F.; Kaplan, L.; Kubš, J. Tectona grandis Linn. and its Fire Characteristics Affected by the Thermal Modification of Wood. *Bioresources* **2017**, *12*, 2805–2817. [CrossRef]
15. Vandličková, M.; Marková, I.; Makovická Osvaldová, L.; Gašpercová, S.; Svetlík, J.; Vraniak, J. Tropical Wood Dusts—Granulometry, Morfology and Ignition Temperature. *Appl. Sci.* **2020**, *10*, 7608. [CrossRef]
16. Antov, P.; Krišťák, L.; Réh, R.; Savov, V.; Papadopoulos, A.N. Eco-Friendly Fiberboard Panels from Recycled Fibers Bonded with Calcium Lignosulfonate. *Polymers* **2021**, *13*, 639. [CrossRef]
17. Dukarska, D.; Pedzik, M.; Rogozinska, W.; Rogozinski, T.; Czarnecki, R. Characteristics of straw particles of selected grain species purposed for the production of lignocellulose particleboards. *Part. Sci. Technol.* **2021**, *39*, 213–222. [CrossRef]
18. Antov, P.; Jivkov, V.; Savov, V.; Simeonova, R.; Yavorov, N. Structural Application of Eco-Friendly Composites from Recycled Wood Fibres Bonded with Magnesium Lignosulfonate. *Appl. Sci.* **2020**, *10*, 7526. [CrossRef]
19. Taghiyari, H.R.; Hosseini, S.B.; Ghahri, S.; Ghofrani, M.; Papadopoulos, A.N. Formaldehyde Emission in Micron-Sized Wollastonite-Treated Plywood Bonded with Soy Flour and Urea-Formaldehyde Resin. *Appl. Sci.* **2020**, *10*, 6709. [CrossRef]
20. Aristri, M.A.; Lubis, M.A.R.; Yadav, S.M.; Antov, P.; Papadopoulos, A.N.; Pizzi, A.; Fatriasari, W.; Ismayati, M.; Iswanto, A.H. Recent Developments in Lignin- and Tannin-Based Non-Isocyanate Polyurethane Resins for Wood Adhesives—A Review. *Appl. Sci.* **2021**, *11*, 4242. [CrossRef]
21. Papadopoulos, A.N. Advances in Wood Composites III. *Polymers* **2021**, *13*, 163. [CrossRef] [PubMed]

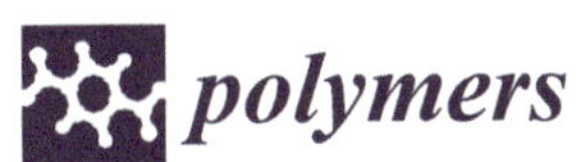

Article

Weathering Stability and Durability of Birch Plywood Modified with Different Molecular Weight Phenol-Formaldehyde Oligomers

Juris Grinins [1,*], Vladimirs Biziks [2], Brendan Nicholas Marais [2], Janis Rizikovs [1] and Holger Militz [2]

1   Latvian State Institute of Wood Chemistry, 27 Dzerbenes Str., LV-1006 Riga, Latvia; janis.rizikovs@gmail.com
2   Georg-August University of Goettingen, Wood Biology and Wood Products, Büsgenweg 4, 437077 Göttingen, Germany; vbiziks@gwdg.de (V.B.); bmarais@uni-goettingen.de (B.N.M.); hmilitz@gwdg.de (H.M.)
*   Correspondence: jurisgrinins@inbox.lv

Abstract: This study investigated the effect of phenol-formaldehyde (PF) resin treatment on the weathering stability and biological durability of birch plywood. Silver birch (*Betula pendula*) veneers were vacuum-pressure impregnated with four different PF resins with average molecular weights (M_w) of 292 (resin A), 528 (resin B), 703 (resin C), and 884 g/mol (resin D). The aging properties of PF resin modified birch plywood were analyzed using artificial weathering with ultraviolet (UV) light, UV and water spray, and weathering under outdoor conditions. The same combinations of PF-treated plywood specimens were then tested in soil-bed tests to determine their resistance against soft-rot wood decay. It was not possible to compare weathering processes under artificial conditions to processes under outdoor conditions. However, the weathering stability of birch plywood treated with PF resins A, B, and C, scored better than plywood treated with commercial resin D (regardless of solid content concentration [%]). Results from unsterile soil bed tests showed improvements in resistance to soft-rot wood decay compared to untreated plywood and solid wood. Mass loss [%] was lowest for birch plywood specimens treated with resin of highest solid content concentration (resin D, 20%). Provisional durability ratings delivered durability class (DC) ratings of 2–3, considerably improved over untreated solid wood and untreated birch plywood (DC 5).

Keywords: birch plywood; molecular weight; phenol-formaldehyde resin; soft-rot; weathering stability

Citation: Grinins, J.; Biziks, V.; Marais, B.N.; Rizikovs, J.; Militz, H. Weathering Stability and Durability of Birch Plywood Modified with Different Molecular Weight Phenol-Formaldehyde Oligomers. *Polymers* **2021**, *13*, 175. https://doi.org/10.3390/polym13020175

Received: 11 December 2020
Accepted: 30 December 2020
Published: 6 January 2021

Publisher's Note: MDPI stays neutral with regard to jurisdictional claims in published maps and institutional affiliations.

1. Introduction

Wood is increasingly being used for outdoor applications, yet it is still limited by complex wood-water-ultraviolet (UV) light interactions. During weathering, wood is cycling through wet and dry states, thus inducing repeated swelling and shrinkage and generating tension stresses. Wood responds to these wetting-drying stresses through creeping and surface cracking. Once the stresses exceed the fracture strength of wood, it has a tendency to develop longer and deeper cracks at later stages [1]. These cyclic changes in moisture content and dimensions are most pronounced at the wood surface, which is directly exposed to rain, humidity and sunlight (UV and visible light). Moreover, UV light exposure intensifies crack formation because photodegradation weakens the wood surface and degrades its microstructure [2]. The presence of cracks and other defects is thus a major drawback and may reduce its service life, market value and mechanical strength. These defects also lead to increased water uptake, thus producing optimal moisture conditions for wood-decay fungi to attack [3].

Silver birch (*Betula pendula*) plywood is widely used in construction, interior and exterior decoration, vehicle construction, sports equipment, furniture, packaging materials, and toy production. However, its application in outdoor, high humidity conditions is limited due to poor biological durability (durability class 5 according to EN 350-1:2016 [4]).

The effect of wood degrading fungi, humidity and UV light, essentially impair the technical properties of plywood through weakening of the bonded veneers, which in-turn weaken mechanical strength and deteriorate the surface finish. Moisture easily penetrates into veneer layers, which causes swelling and characteristic waves on the plywood surface. Water uptake potential can be decreased by covering the plywood with a hydrophobic (water repelling) laminate. Melamine and phenolic resins are mostly used for the production of resin laminates, which are subsequently hot-pressed onto one or both sides of the finished plywood board surfaces. Coating plywood protects the top veneer layer from direct contact with water and UV light, but when the upper laminate coating is damaged, the inner veneer layers can swell, provoking noticeable surface failure and promoting attack by biodegradation agents. Most wood species used for plywood production have poor resistance to swelling, biodegradation and UV weathering under outdoor and high humidity conditions. Therefore, when using plywood in conditions with high humidity, where regular wetting is possible, it is necessary not only to cover it from the outside with a hydrophobic laminate, but also to treat the single veneers constituting the entire plywood board. Prolonging the service life of wood and wood-based products results in a positive effect on greenhouse gas emissions by storing biomass carbon for longer periods [5].

Wood modification alters the material properties to a greater extent than preservative treatment, and the magnitude of changes depends on the applied modification method [6]. Wood modification can simultaneously overcome several weaknesses of wood, such as poor dimensional stability, low decay resistance, high equilibrium moisture content, and aesthetical issues such as optical appearance can be diversified and enhanced. Wood impregnated with most thermosetting resins causes changes in colour [7]. Compared to untreated wood, acetylated wood [8,9], glutaraldehyde treated wood [10], DMDHEU- and melamine- treated wood [7] exposed to accelerated or outdoor weathering develops fewer cracks because of its improved dimensional stability. Surface discoloration (graying) and crack development during longer exposure times is reduced, whereas in the case of thermal treatment, the rate of graying and crack development is the same or even faster than that of untreated wood [3,11]. In contrast, the phenol formaldehyde (PF) treated boards remained darker, ranging from light brown to dark brown. PF resin turns wood red-brown, due to differences in pH, because wood is acidic and resole PF resins are alkaline [12]. This change in optical appearance depends on both the resin type and the average molecular size of the PF resin oligomers used for treatment. PF resin acts as a UV absorber for the photostabilization of wood [13]. PF resin is also an antioxidant, which may also impact its ability to photostabilize wood [14]. Kielmann and Mai [15,16] have concluded that the surface of PF-treated wood without a coating has improved resistance against photodegradation compared to the surface of N-methylol melamine (NMM)-treated wood because PF inhibits lignin degradation. The resistance of wood treated with low molecular weight PF resin to weathering can be improved by increasing the concentration of PF resin and by combining it with a water soluble hindered amine light stabilizer [17]. PF resins could be modified with ferric chloride and a mixture of ferric sulphate and hydrolysable polyphenols to darken the colour of European beech wood (*Fagus sylvatica* L.) and enhance colour stability [18]. Although a proper improvement in dimensional stability and biological durability and a considerable reduction in water sorption are attained, the appearance of the treated wood still undergoes considerable changes during weathering. Therefore, over the past decade, the combined approach of coating chemically treated wood has become an increasing point of interest as a way of increasing wood service life.

As mentioned, outdoor wooden components are subjected to a variety of biotic and abiotic degradation factors. Additionally, wood used in soil contact is of particular risk due to the permanent to semi-permanent presence of moisture [19]. Important considerations for the successful proliferation of wood-decaying fungi include a carbon substrate, moisture, temperature, and oxygen [20]. Various wood decaying fungi; brown-, white-, and soft-rot fungi, can all be found on wood utilized in-ground, but these decay types can vary significantly, not only in frequency and spatial distribution, but also in combinations from one site to the next [21,22], and with decay progress [23]. Soft-rot seems to be able to cope with high soil moisture content (MC_{soil}) better than brown- and white-rot fungi, and continues to remain active over a broader temperature range (T_{soil}) compared to brown- and white-rot fungi [21,24,25].

In the pursuit of new techniques for improved wood protection, a rapid assessment of the technique's effectiveness can be attained through laboratory tests. Such tests deliver results quickly and thus form the basis for the decision for further test steps. It is of the greatest interest that the predictions obtained from these tests can be transferred to practice (field tests) with a high degree of certainty. One possibility to better assess the risk factors that determine wood degradation when used in contact with soil is to test the wood in-field and in contact with soil. This test in an option when assessing the behavior of wood preservatives by DIN EN 252:2015 [26]. However, depending on the duration and characteristics of the vegetation periods, it may take several years before results from this type of test are available. Since this method requires a lot of time, the use of pure-culture basidiomycete tests such as CEN/TS 15083-1:2005 [27] and unsterile soil bed tests such as CEN/TS 15083-2:2005 [28] under controlled laboratory conditions are often employed. The results from these tests are designed to complement each other in combination with longer-term field tests using specimens of larger dimension. Newly developed wood preservative and modification techniques can therefore be tested to deliver preliminary durability ratings.

This study compared the color changes of a developed birch plywood modified with different PF resins after weathering under artificial conditions with UV light only, and UV light and water spraying, and under real outdoor conditions. Additionally, so called terrestrial microcosms (TMC) consisting of unsterile organic soil were used to test the resistance of the developed birch plywood against wood decay by soft-rot fungi. This study impregnated birch wood veneers with PF resin solutions of different solid content concentrations and PF oligomer sizes, to understand the effect on dimensional stability, weathering performance, and biological durability. Theoretically, the photostability and resistance to wood decaying fungi of the wood material treated with low molecular weight PF resins should be improved. Also, increasing the concentration of PF resin should improve the weathering and biological durability of the treated wooden material. Criteria for birch veneer treatment was based on minimum WPG requirements to achieve the maximum improvement in properties, therefore to limit PF resin load in the wood material.

2. Materials and Methods

2.1. Weathering Stability Test

2.1.1. PF Resin Synthesis

For the synthesis of resin A, B, and C, phenol was hydroxymethylated under alkaline reaction conditions, whereby the molar ratio of formaldehyde/phenol/sodium hydroxide was 2.0/1.0/0.2. During the synthesis of each resin, a measured amount of phenol and aqueous sodium hydroxide solution was weighed out in a 4-neck laboratory reactor (1 L) equipped with a thermometer, dropping funnel, reflux condenser and stirrer. Some ethanol was also added in order to maintain a homogeneous reaction. The 4-neck reactor was submerged in a thermostatic water-bath. As soon as the temperature in the flask reached the necessary synthesis temperature, the aqueous formaldehyde solution was added slowly via a drip over a 25–30 min period. The reaction temperature (65, 75, 85 °C) was kept constant during the entire reaction period (2 and 4 h). The resol synthesis was ended by cooling the reactor with cold running water and allowing the resol to cool down to 20 ± 3 °C. Resin D was provided by Prefere Resins Germany GmbH (Erkner, Germany). The different characteristic parameters of the synthesized PF resins are listed in Table 1.

Table 1. Characteristic parameters of PF resins used in the study.

Resin	Viscosity (mPas)	Solid Content (%)	M_n (g/mol)	M_w (g/mol)	Dispersity $Q = (M_w/M_n)$	Free Formaldehyde (%)	pH
A	75	49.4	220	292	1.327	0.6	10.0
B	125	50.0	338	528	1.562	0.4	10.3
C	282	49.7	467	884	1.892	0.5	10.4
D	216	55.9	414	703	1.698	<0.8	9.4

2.1.2. Resin Characterisation

The dynamic viscosity of liquid PF resins was determined by a Fungilab Viscolead Adv meter (Fungilab S.A., Barcelona, Spain) equipped with a suitable spindle. The non-volatiles content (solid content) was determined according to DIN EN ISO 3251:2019 [29]. The pH value was determined using a digital pH meter (GPH 114 Greisinger, Regenstauf, Germany) by inserting the pH meter electrode into the PF resins. The pH meter was calibrated with buffer solutions at pH 4.0 and 10.0 prior pH measurements. Free formaldehyde content was determined by the hydroxylamine hydrochloride method according to DIN EN ISO 9397:1997 [30].

For gel permeation chromatography (GPC) analysis, a 1260 Infinity system (degasser, isocratic pump, automatic liquid sampler, heatable column compartment, RID, MWD @ 280 nm, Agilent (Santa Clara, CA, USA) was used, where: column: $3 \times$ PLgel $5\,\mu$ (50 Å, 100 Å, 1000 Å), 7.5×300 mm; solvent: tetrahydrofuran (THF); flow rate: 0.6 mL/min; flow rate marker: acetone; calibration: polystyrene standard.

Approximately 40 mg of resin was dissolved in 5 mL of THF. If the resin did not completely dissolve, it was sonicated with slow addition of H_2SO_4 (5% in methanol) until neutral. If the resin was dissolved, but precipitate from additives (such as salts) remained, the mixture was filtered with a syringe filter.

2.1.3. Treatment of Veneer Material

Air-dried veneer sheets of silver birch wood ($300 \times 300 \times 1.5$ mm^3 and $400 \times 400 \times 1.5$ mm^3 L $\times$ R $\times$ T) were prepared for impregnation. Oven-dry mass was determined after drying at 103 ± 2 °C for 24 h. Previous experiments suggested that treatment of birch wood with commercial PF resin solutions of 10% concentration delivered weight percentage gain (WPG) of 9–12%, which subsequently improved dimensional stability and allowed for a provisional durability class (DC) rating of 1 against decay fungi [31,32]. Veneers ($300 \times 300 \times 1.5$ mm^3) were conditioned to 5–6% moisture content and impregnated with 10% solid content concentration solutions of PF resins A, B, and C. Impregnation was carried in a 340-litre chamber produced by Wood Treatment Technology (WTT, Grindsted, Denmark). Veneers (n = 40) were placed in a tub filled with the resin solution. The veneers were prevented from floating using a mesh grid and heavy weight. Impregnation was carried out in two steps. The first, vacuum step (1 h, 0.1 bar of pressure), was used to ensure the free air was purged from the specimens. The chamber was then pressurized to ensure sufficient diffusion of the PF oligomers into the wood cell walls (1 h, 4 bars of pressure). The veneers were then removed from the impregnation chamber and measured for solution uptake. The remaining resin solution was drained from the tub and the veneers were positioned vertically to allow excess resin solution to drip off. After impregnation, veneers were oven dried to 4–6% moisture content, with moderate air circulation and air exchange for 72 h using incrementally rising temperature intervals from 30–50 °C. A subset of PF resin impregnated veneers from each resin impregnation treatment was measured for WPG. These veneers were cured at 140 °C for 1 h. The WPG was calculated to assess the amount of PF resin in the veneers. The average WPG was calculated for each treatment according to Equation (1) below:

$$\text{WPG} = \frac{(M_1 - M_2)}{M_2} \times 100 \tag{1}$$

where:

- WPG is the weight percentage gain [%];
- M_1 is the oven-dried mass of the modified wood specimens [g];
- M_2 is the oven dry mass of the unmodified wood specimens [g].

Commercial resin D was used to evaluate the behaviour of different loading of PF resin in veneers. Therefore, before being used for impregnation, the stock solution of the PF resins D was diluted with distilled water to 10, 15 and 20% solid content concentration. Veneers ($400 \times 400 \times 1.5$ mm^3) were conditioned to 5–6% moisture content and impregnated with PF resin solutions in a 1000-litre impregnation chamber at the University of Göttingen. The impregnation and drying parameters were set the same as for veneers of $300 \times 300 \times 1.5$ mm^3. The WPG was calculated to assess the amount of PF resin in the veneers. Veneer WPG after impregnation and curing for both veneer dimensions is shown in Table 2 below.

Table 2. Veneer WPG after treatment with PF resin solutions.

Resin Treatment	A 10%	B 10%	C 10%	D 10%	D 15%	D 20%
WPG (%)	14.6 ± 1.8	13.2 ± 1.6	13.9 ± 1.5	12.5 ± 1.0	19.9 ± 3.3	27.5 ± 1.8

2.1.4. Plywood Production

Standard PF adhesive (sourced from plywood industry partners) was applied to the veneer sheets in preparation for pressing into plywood (nine layers). PF adhesive viscosity was 380 mPas, solid content 44.5%, free formaldehyde content <1% and pH 12.6. A quantity of 150 g/m^2 was applied to one surface of eight of the nine plywood sheets constituting a 9-layer plywood board. After adhesive application, veneers were pre-dried at room temperature for 15 min (adhesive open time) before assembling nine individual veneer sheets in a crosswise, perpendicular fashion in preparation for pressing. Assembled sheets were then pressed in a hot press (Joos LAP 40, Gottfried Joos Maschinenfabrik GmbH & Co. KG, Pfalzgrafenweiler, Germany) at 140 °C and 1.5 N/mm^2 for 20 min (90 s/mm) to deliver a plywood board with thickness of approximately 11 mm. Thereafter specimens were prepared for artificial weathering, outdoor weathering and unsterile soil-bed tests with dimensions of $150 \times 70 \times 11$ mm^3, $110 \times 40 \times 11$ mm^3 and $100 \times 10 \times 11$ mm^3, respectively.

2.1.5. Artificial Weathering Tests

Artificial weathering tests were performed in a QUV accelerated weathering tester, (Q-Lab Europe, Ltd., Farnworth Bolton, England) equipped with UVA-340 type fluorescent lamps. Three plywood specimens ($150 \times 70 \times 11$ mm^3) were used for each resin treatment. The lamps provided a good simulation of sunlight in the short wavelength region; from 295 nm to 365 nm, with a peak emission at 340 nm. The UV radiation flux density at 340 nm was 0.89 W/m^2 and the chamber temperature throughout the test was kept constant at 60 °C. The intensity of the full UV spectrum's (290–400 nm) irradiation was 21.5 W/m^2. In the study, two different artificial weathering tests were carried out. The first test involved only UV irradiation. This test was regularly suspended to measure the change in colour of the specimens. The total duration of the test was 360 h. The second artificial weathering test involved both UV irradiation and water spray. The test involved the following steps; 2.5 h of UV radiation at the same conditions as described earlier, followed by 30 min of water spray. In total, 60 cycles were preformed to reach an exposure time of 180 h from which 150 h accounted for UV irradiation. Colour measurements after both weathering methods was performed.

2.1.6. Surface Colour Measurements

Colour of the specimens was measured with a CM-2500d spectrophotometer (Konica Minolta, Ramsey, NJ 07446, USA) and expressed according to the CIELAB three-dimensional colour system. On each of the specimen, five locations were randomly chosen and marked. For the marked locations, the colour was measured before and after the weathering tests as well as during the test after 2, 4, 8, 16, 24, 48, 120, 192, 264 and 360 h. The colour was measured to evaluate the discolouration caused by weathering. The total colour change ΔEab was calculated according to the Equation (2) below. L^* is a lightness parameter, a^* is a chromaticity parameter which represents red-green coordinates and b^* is a chromaticity parameter which represents yellow-blue coordinates:

$$\Delta Eab = \sqrt{(L_x^* - L_0^*)^2 + (a_x^* - a_0^*)^2 + (b_x^* - b_0^*)^2} \tag{2}$$

where:

- L_0^*, a_0^*, b_0^* is the value on coordinate axis for the specific parameter at the beginning;
- L_x^*, a_x^*, b_x^* is value on coordinate axis for the specific parameter after weathering.

2.1.7. Outdoor Weathering and Fungal Tests

Six replicates of all PF resin treated plywood with dimensions of $110 \times 40 \times 11$ mm^3, along with untreated specimens, were used. For half of the specimens (3 from each impregnation treatment), the side edges were coated with urethane alkyd paint (brushed on application, three coats). According to EN 152:2011 [33], the samples were placed in an aluminium rack at 45°, one meter above the ground, facing the south direction with most severe weather conditions. Microorganisms were allowed to attack the specimens. The test site in the courtyard of Latvian State Institute of Wood Chemistry (Riga, Latvia) was free of vegetation, shade and extreme environmental conditions. The test lasted for 3 months, from 19 June to 21 September 2020 and weather data are listed in Table 3.

Table 3. Weather data for the outdoor, aboveground weathering test site for the 3-month test period obtained from ©weatheronline.co.uk.

Month	Mean Rainfall [mm]	Min–Max Temperature Range [°C]	Relative Humidity Range [%]	UV Index Range
June 2020	5	15–30	50–80	6–7
July 2020	6	10–28	55–90	5–6
August 2020	12	11–28	55–90	4–5
September 2020	12	7–24	55–90	2–4

Mould and blue stain growth was evaluated once per month by stereomicroscopy (M8, Leica, Wetzlar, Germany) and digital photography (2 MB) according to the rating scale 0–4: 0—clean, 0% attack; 1—trace, $\leq$5% growth; 2—slight, 6–25% growth; 3—medium, 26–50% growth; 4—severe, >50% growth. Colour change measurements were also performed once per month.

2.2. Unsterile Soil-Bed Test: Resistance against Soft-Rot Wood Decay

Terrestrial microcosms (TMCs) in accordance with CEN/TS 15083-2:2005 were utilised to test the resistance of the developed plywood material against soft-rot wood decay. The standard stipulates that a natural topsoil or a fertile loam-based horticultural soil substrate is used, with pH 6–8, and no additives. The soil should have a WHC$_{soil}$ of 20–60%, MC$_{soil}$ equal to 95%WHC$_{soil}$, and the test should be conducted in a dark, climate-controlled room set to a temperature of 27 °C and relative humidity of 65%.

2.2.1. Soil Substrate

The soil substrate was a horticultural compost produced at the forest botanical garden at the University of Göttingen's North Campus. The compost comprised of fallen leaves and cuttings from grass and trees. Soil was passed through a sieve with nominal aperture size of 8.5 mm. WHC_{soil} was then determined according to the 'cylinder sand bath method' according to ISO 11268-2:2012 [34]. The test deviated from the standard in that no silica sand was added to the soil in order to reduce the soil's water-holding capacity. Silica sand acts to standardize and reduce the soil's inoculum potential to attain reproduceable wood decay results which serves as a provisional durability rating.

2.2.2. Determination of the Soil-Water Holding Capacity (WHC_{soil})

Soil was inserted into polyethylene cylinders 10 cm long with 4 cm diameter. The bottoms of the cylinders were covered with a fine polymer grid and filter paper (MN 640 W 70 mm). All cylinders were filled with soil to a height of 5–7 cm and saturated in an 8 cm high water bath for 3 h. After the saturation period, the cylinders were placed on a water saturated sand bath for 2 h to allow unbound water within the soil-filled cylinders to drain to reach the equivalent of field capacity. The soil samples were then weighed wet, as well as after oven-drying at 103 $\pm$ 2 °C for 24 h. WHC_{soil} [%] was calculated according to Equation (3) below. The compost soil batch deliver WHC_{soil} of 105%:

$$WHC_{soil} = \left(\frac{m_s - m_0}{m_0} \right) \times 100 \tag{3}$$

where:

- WHC_{soil} is the soil water-holding capacity [%];
- m_s is the saturated soil mass [g];
- m_0 is the oven-dry soil mass [g].

2.2.3. Determination of the Soil Moisture Content (MC_{soil})

In order to ensure an equal quantity of soil was used across all TMC boxes, measurements of soil moisture content (MC_{soil}) were used. The soil quantity decided on for all TMC boxes was based on the based on the oven-dry mass of the soil [g], and was dependent on the box's dimension. Each TMC box was filled to ensure a soil height of approximately 12 cm, to this end the oven-dry soil mass decided on for all TMC boxes amounted to 6000 g. The plastic TMC boxes were initially weighed in dry, empty state. Then, TMCs were over-filled with soil (i.e., >12 cm in height), with the box (including soil) weighed again to calculate the mass [g] of the soil component added to the box. Five MC_{soil} samples of 50 g each were then taken from multiple locations throughout the TMC box to ensure an average, but homogenously distributed MC_{soil} measurement was attained. Soil samples were weighed to the nearest 0.001 g, oven-dried at 103 °C for 24 h, and weighed again. MC_{soil} was calculated according to Equation (4) below. Once a representative MC_{soil} measurement was attained, a rearrangement of Equation (4) below was carried out to calculate the quantity of 'wet soil' (m_w) required to be removed from the TMC box to attain 6000 g of oven-dry soil (m_0):

$$MC_{soil} = \left(\frac{m_w - m_0}{m_0} \right) \times 100 \tag{4}$$

where:

- MC_{soil} is the soil moisture content [%];
- m_w is the wet soil mass [g];
- m_0 is the oven-dry soil mass [g].

2.2.4. Preparation of Soil Substrates to Reach Target Soil Moisture Content ($MC_{soil,target}$)

Once 6000 g of oven-dry soil was filled into each of the six TMC boxes, MC_{soil} for each TMC was set to equal 95% of the measured WHC_{soil}, expressed as 95%WHC_{soil}. In order to understand the quantity of distilled water required to reach the MC_{soil} equal to 95%WHC_{soil}, a target soil moisture metric $MC_{soil,target}$ [%] was defined. Equation (5) below was used to calculate the mass [g] in distilled water required to add to the soil mixture to reach $MC_{soil,target}$. Distilled water was subsequently added to each of the TMCs to reach $MC_{soil,target}$. To account for losses in MC_{soil} resulting from fungal activity and evaporation, rewetting to $MC_{soil,target}$ occurred once per week throughout the 24-week incubation period:

$$m_{water} = \left(\frac{MC_{soil,target} - MC_{soil,current}}{100} \right) \times m_{total,dry} \tag{5}$$

where:

- m_{water} is the mass of distilled water to add to the soil mixture [g];
- $MC_{soil,target}$ is the target soil moisture content [%];
- $MC_{soil,current}$ is the current moisture content of the soil mixture before adding any additional water [%];
- $m_{total,dry}$ is the oven-dry mass of the total soil mixture [g].

2.2.5. Preparation and Exposure of Wood Specimens

As already mentioned, some aspects of this TMC test against soft-rot wood decay deviated from the standard CEN/TS 15083-2:2005 [28]. One deviation was the inoculum aggressiveness of the soil material used, the other deviation, which also influenced the decision to use a more aggressive soil was the specimen dimension. Plywood boards of birch with nine veneer layers were prepared. The final height (thickness) of the 9-layer plywood boards amounted to 11.5 mm. Thereafter, individual specimens were prepared from the larger plywood boards to final dimensions of $10 \times 11.5 \times 100$ mm^3.

Before specimens were prepared from the larger plywood boards, the boards were conditioned to wood moisture content (MC_{wood}) of $12 \pm 2\%$. MC_{wood} was tested using a 2-pronged electrical resistance moisture content measuring device. Specimens were then prepared from strips of the boards, with a cross-section of 10 ± 0.1 mm $\times 11.5 \pm 0.1$ mm (board thickness). Transverse cuts of the cross-section delivered sharp edges and a fine-sawn finish to the end-grain surface, with a final specimen length of 100 ± 1 mm. All specimens were free from obvious defects such as cracks, decay and discolouration.

After specimen preparation, all specimens were oven-dried at 103 °C for 24 h and weighed for oven-dry mass to the nearest 0.001 g. Prior to soil exposure, all specimens were again conditioned to MC_{wood} of $12 \pm 2\%$ (confirmed by Equation (6) and buried 4/5 of their length into the soil substrate with 38 specimens per TMC box. For control (plywood, birch and beech solid wood) 30 replicate specimens were used with three specimen removal intervals (16, 20, 24 weeks). For each PF-treated plywood type 20 replicate specimens were used, with two specimen removal intervals (16 and 24 weeks). After soil exposure, specimens were removed, cleaned of remaining soil and again oven-dried at 103 °C for 24 h. Specimens were then weighed again to the nearest 0.001 g with oven-dry wood mass loss (ML_{wood}) calculated according to Equation (7) below. Mean ML_{wood} and standard deviation of mean ML_{wood} was calculated according to Equation (8) and Equation (9) below:

$$MC_{wood} = \left(\frac{m_3 - m_2}{m_2} \right) \times 100 \tag{6}$$

where:

- MC_{wood} is the wood moisture content, [%];
- m_3 is the wood specimen's mass after TMC exposure, [g];

- m_2 is the wood specimen's oven-dry mass after TMC exposure, [g].

Oven-dry mass loss (ML_{wood}) of wood was calculated according to Equation (7) below:

$$ML_{wood} = \left(\frac{m_1 - m_2}{m_1}\right) \times 100 \tag{7}$$

where:

- ML_{wood} is the wood specimen's oven-dry mass loss [%];
- m_1 is the wood specimen's oven-dry mass before TMC exposure [g];
- m_2 is the wood specimen's oven-dry mass after TMC exposure [g].

Mean ML_{wood} was calculated according to Equation (8) below:

$$\text{mean } ML_{wood} = \frac{1}{n}\sum_{i=1}^{n} x_i \tag{8}$$

where:

- mean ML_{wood} is the arithmetic mean of the oven-dry mass loss of the sample population;
- x_i is the oven-dry mass loss (ML_{wood}) of each individual wood specimen in the sample population;
- n is the total number of wood specimens in the sample population.

Standard deviation of mean ML_{wood} was calculated according to Equation (9) below.

$$s = \sqrt{\frac{\sum_{i=1}^{n}\left(x_i - \bar{x}\right)}{n-1}} \tag{9}$$

where:

- s is the standard deviation of the sample population;
- x_i is the oven-dry mass loss (ML_{wood}) of each individual wood specimen in the sample population;
- $\bar{x}$ the mean oven-dry wood mass loss (mean ML_{wood}) of the sample population [g];
- n is the total number of wood specimens in the sample population.

2.2.6. Calculation of x-Value towards Provisional Durability Rating

According to CEN/TS 15083-2:2005 [28], the percentage oven-dry mass loss of the tested wood specimens is used to determine the resistance of hardwood test timbers to wood decay by soft-rotting fungi. The calculation of the x value based on the median oven-dry mass loss of the treated test specimens and the untreated reference specimens is used to define a provisional durability class (DC) rating. However, DC ratings do not equal use class ratings (like those covered in EN 335:2013 [35]. Use classes, and the combination of use classes and DC to assess wood suitability for a particular application are addressed in EN 460:1994 [36] and prEN 460:2019 [37], respectively. Equation (10) below was used to calculate the x value:

$$x = \frac{\text{median } ML_{wood} \text{ of the various treated test specimen group}}{\text{median } ML_{wood} \text{ of reference test specimens}} \tag{10}$$

where:

- x is the x value used for interpretation in a provisional durability rating scale;
- ML_{wood} is the oven-dry mass loss of the relevant wood specimen.

3. Results and Discussion

3.1. Artificial Weathering (UV Only)

There is a demand among end users for new plywood products with predictable, long-term aesthetic properties. Plywood products with improved colour stability naturally have an advantage over competitors. In order to evaluate how UV radiation affects the PF resin treated plywood developed in this study, colour parameters were evaluated at different exposure intervals. The results presented in Figure 1 show the colour parameters of PF treated and untreated birch plywood after UV weathering of 360 h. Both the untreated controls and the plywood treated with PF resin became darker after UV weathering (decreasing ΔL^*), and this effect was more pronounced for plywood treated with commercial resin D (M_w = 703 g/mol) at all tested concentrations (Figure 1a). This is also clearly seen in the specimen photos after UV weathering (Figure 4: UV only).

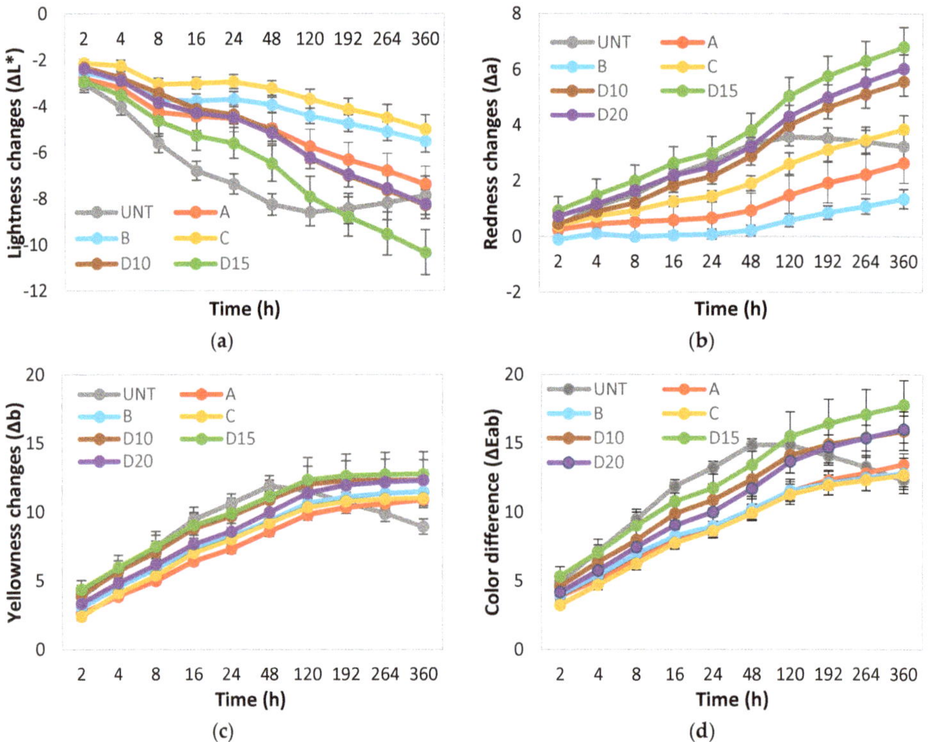

Figure 1. Changes in colour of untreated (UNT) and PF resin treated plywood after exposure to UV light during 360 h in QUV camera: (**a**) Changes in the CIE parameter ΔL^* (lightness); (**b**) Changes in the CIE parameter Δa (red-green); (**c**) Changes in the CIE parameter Δb (yellow-blue); (**d**) colour difference parameter ΔEab.

Untreated and PF resin treated plywood became redder during UV weathering (increasing Δa). The change in red colour parameter was minimal for resin B (M_w = 528 g/mol) while resin A and B (M_w = 292 and 884 g/mol) had similar changes compared to the untreated plywood. Resin D (M_w = 703 g/mol) treated plywood at all tested concentrations had the highest changes in redness parameter (Figure 1b).

For all PF treated specimens, the yellowness (Δb) parameter increased similarly to between 11 and 13 units after 360 h, while untreated plywood became less yellow after UV weathering reaching 9 units after 360 h of exposure (Figure 1c).

The total colour change (Figure 1d: ΔEab) for all PF treated plywood specimens was between 12 and 18 units after 360 h of exposure. Resins A, B and C (M_w = 292, 528 and 884 g/mol) had lower ΔEab values (12–13 units) compared to commercial resin D (M_w = 703 g/mol), at all tested concentrations (15–18 units). Resin D-treated plywood did not show major differences in colour change between different concentrations, D15 showing the highest colour difference. For all PF-treated specimens, colour changes under the influence of UV light occured most rapidly in the first 24 h, after which the rate of change slowed down considerably. For untreated plywood, the change in colour parameters reached a maximum during the first 48 h and then decreased slightly further during the testing period.

3.2. Artificial Weathering (UV + Water Spray)

In order to recreate so called "real-life" conditions, artificial weathering using a combination of UV radiation and moisture spraying was used. Moisture spraying recreated the effect of rain, mist and dew. Results presented in Figure 2 show colour parameters of PF treated and untreated birch plywood after UV weathering and water spraying, after a total test time of 1000 h.

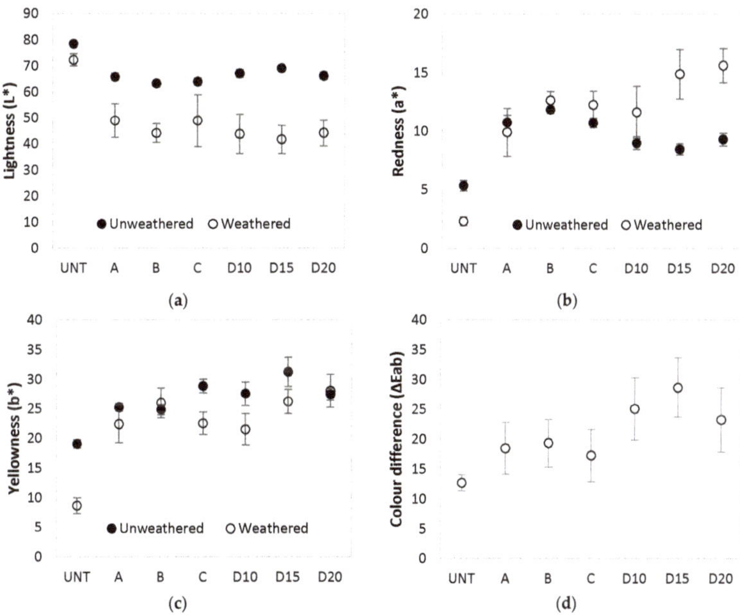

Figure 2. Changes in colour of untreated (UNT) and PF resin treated plywood after exposure to UV light and water spray after 1000 h in QUV camera: (**a**) Changes in the CIE parameter L* (lightness); (**b**) Changes in the CIE parameter a* (red-green); (**c**) Changes in the CIE parameter b* (yellow-blue); (**d**) Colour difference parameter ΔEab.

After 100 h of UV weathering and water spraying, plywood treated with PF resin as well as untreated plywood controls became darker (L* decreased) compared to un-weathered reference specimens. For PF resin-treated specimens, lightness decreased by 15–27 units, but for untreated plywood by only 6 units (Figure 2a). For untreated plywood after UV and water spraying, the redness parameter (a*) decreased by 3 units and yellow-ness (b*) by 10 units (Figure 2b,c). For plywood treated with resins A and B (M_w = 292, 528 g/mol), the redness and yellowness parameters were affected negligibly, while for resin C (M_w = 884 g/mol) only yellowness decreased. For plywood treated with resin D (M_w = 703 g/mol), redness increased by 2–6 units at all concentrations, while yellowness was only affected for 10 and 15% solid content concentration treatments (b* decreased by 5–6 units). The smallest changes in colour (ΔEab) after UV + water spray were observed for untreated plywood (13 units), for resins A, B and C, the changes reached 17–19 units, but the largest changes (23–28 units) were for resin D treated specimens at all tested concentrations (Figure 2d). A similar trend was observed for total colour change after only UV weathering.

Visual assessment of the specimens showed that the untreated plywood turned gray with surface cracking after 1000 h of UV and water spraying. All PF resin treated specimens acquired an uneven brownish colour with some shades of grey that differed considerably from the original colour of the specimens prior to weathering (Figure 4: UV + water spray). As a result of UV light and water spraying, PF resin that was not fixed within the wood cell wall leached to the plywood surface, causing uneven discolouration patterns during simulated weathering. Kielmann and Mai [15] have tested PF treated (25% (*w*/*w*) aqueous solution) beech (*Fagus sylvatica* L.) wood boards with and without coatings. After UV and water spray PF treated beech surface became darker brown, however the total colour difference (ΔEab) for PF treated wood was similar to untreated wood reaching 25–30 units.

Our data after artificial weathering with only UV light and then with UV and water spraying contradicted general information available in literature that suggests treatment with PF resin significantly improves the colour stability of the wood material [9,15–18]. Data obtained in this study shows that the total colour difference of PF treated specimens is similar (in case of only UV weathering) or even greater than that of untreated plywood.

3.3. Weathering under Outdoor Conditions

Real, outdoor conditions test the ability of a material to resist weathering and discolouration processes most accurately. Therefore, untreated and PF resin treated plywood was also tested outdoors. The results presented in Figure 3 show colour parameters of PF resin treated and untreated birch plywood after 3 months of weathering in real, outdoor conditions. Here, both untreated and PF resin-treated plywood became darker (L* decreased) after weathering. For PF-treated specimens, lightness decreased by 5–17 units, while for untreated plywood, lightness decreased by 23 units (Figure 3a). For untreated plywood after outdoor exposure, the redness parameter (a*) decreased by 4 units and yellowness (b*) by 13 units, with an overall grey discolouration. For plywood specimens treated with PF resins A, B and C (M_w = 292, 528 and 884 g/mol), redness decreased by 3–5 units and yellowness by 5–6 units. For all specimens treated with resin D (M_w = 703 g/mol), the opposite tendency was observed and redness increase was 1–3 units while yellowness 2–5 units (Figure 3b,c). As a result, the largest changes in colour (ΔEab) after outdoor weathering were found for untreated plywood (26 units). For resins A, B and C, the colour changes reached approximately 10 units, but for specimens treated with resin D at all concentrations, the changes reached 12–18 units (Figure 3d). Comparable results were obtained by Evans et al. [17], who exposed PF treated radiata pine veneers in outdoor weathering for 2000 h. They concluded that higher resin loading in wood causes higher total colour changes after weathering. Only 10% of PF solution treated specimens showed lower total colour difference (ΔEab) compared to untreated wood while 20 and 30% treatments had similar values.

Under outdoor weathering conditions, PF resin-treated plywood specimens showed considerably better colour stability than untreated plywood compared to artificial conditions. This suggests that artificial weathering data do not reflect the actual properties of the material in real-use conditions.

After three months of outdoor exposure, untreated plywood specimens turned grey with surface cracking as well as mould and blue stain fungal growth (Figure 4: Outdoor weathering). Specimens treated with PF resins A, B and C also show cracks, mould and blue stain on the surface after outdoor exposure. However, their intensity was lower and the total colour change was less pronounced. Resin D treated specimens (all solid content concentrations) changed colour the most. Mould and blue stain fungal growth developed within surface cracks.

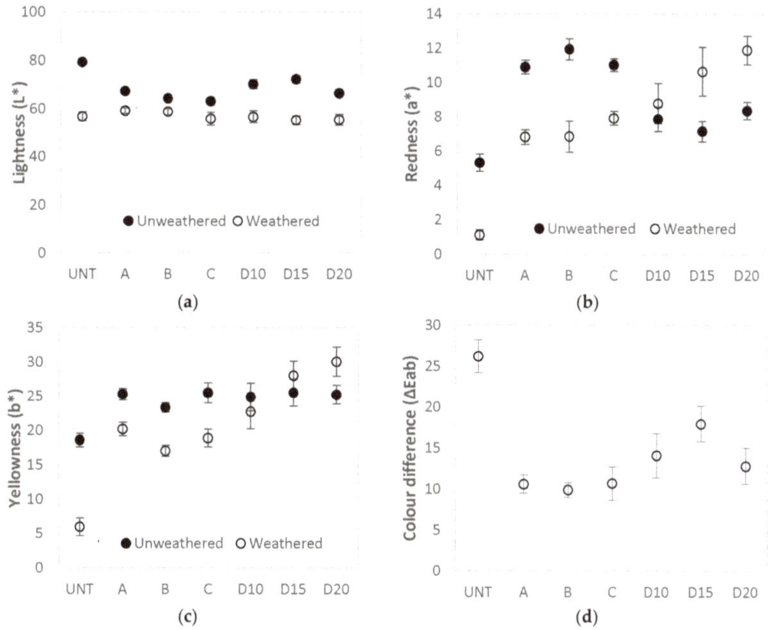

Figure 3. Changes in colour of untreated (UNT) and PF resin treated plywood after exposure to outdoor conditions for 3 months: (**a**) Changes in the CIE parameter L* (lightness); (**b**) Changes in the CIE parameter a* (red-green); (**c**) Changes in the CIE parameter b* (yellow-blue); (**d**) Colour difference parameter ΔEab.

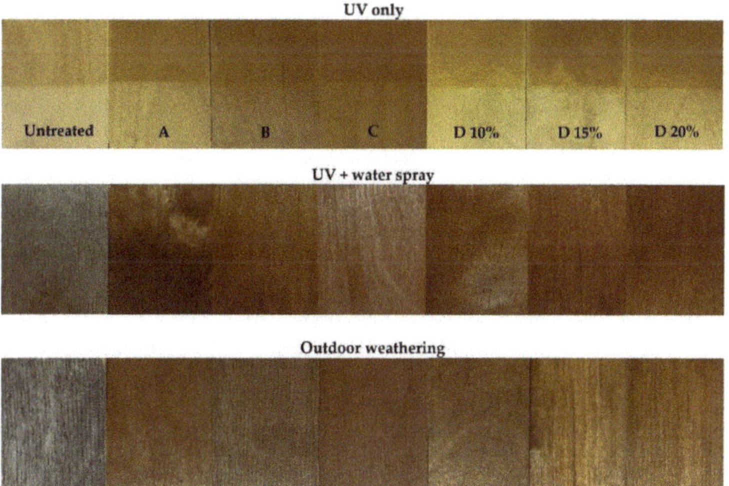

Figure 4. Untreated and PF resin treated plywood photo fixation after weathering with UV only, UV + water spray and in outdoor conditions for 3 months.

3.4. Growth of Wood Discolouring Fungi on PF Treated Plywood

In outdoor aboveground conditions, wood materials are affected not only by abiotic natural weathering through UV radiation and moisture, but also though biotic damage caused by fungi and mould growth. Other anthropogenic sources, such as air pollution, can also deteriorate the surface appearance of wood. Table 4 below shows the growth rate of mould and blue stain fungi on PF resin treated plywood and untreated reference plywood throughout the 3-month outdoor exposure period. After one month, no fungal growth was observed on tested specimens. The first signs of fungal growth on PF treated plywood appeared after 2 months (rated 1–3) of outdoor exposure. After 2 months more than 50% of the untreated plywood surfaces were colonised by fungi (rated 4). After 3 months all PF resin D (M_w = 703 g/mol) treated specimens reached rating 4 (severe growth >50%). Higher resin loading for 15 and 20% treated specimens seemed to show no protective effect against mould and blue stain on the plywood surface.

Table 4. Mould and blue stain growth marks on PF resin treated plywood in outdoor, aboveground conditions tested according to EN 152:2011 [33].

Time (Months)		Resin Treatment						
		Untreated	A	B	C	D 10%	D 15%	D 20%
Coated edges	1	0.0	0.0	0.0	0.0	0.0	0.0	0.0
	2	4.0	3.0	2.0	2.0	2.0	2.0	2.0
	3	4.0	4.0	3.7	3.7	4.0	4.0	4.0
Uncoated edges	1	0.0	0.0	0.0	0.0	0.0	0.0	0.0
	2	4.0	2.0	1.0	2.0	2.0	2.0	2.0
	3	4.0	3.3	3.3	3.0	4.0	4.0	4.0

All PF-treated specimens reached a rating of 3–4 (medium growth 26–50% to severe growth >50%) after 3 months. However, the best results were shown by specimens impregnated with resins B and C (M_w = 528 and 884 g/mol), for which the colour marks rated 3–3.7.

Specimens treated with resin A, B and C (M_w = 292, 528 and 884 g/mol) with coated edges showed greater mould and blue stain growth after 2 and 3 months. Such a tendency was not observed with the specimens treated with resin D at all tested concentrations. In our opinion, this is due to the fact that the specimens with coated edges after wetting under the influence of rain dried up much slower, thus creating a more favourable environment for the development of mould and blue stain. However, the moisture content in specimens during outdoor test was not measured and it is difficult to confirm this assertion. Our data testify that resins A, B and C are relatively well penetrated and fixed in the birch wood cell wall and WPG after leaching decreases by 1.5–2.0% (unpublished results). Maybe this might be attributed to different unfixed PF resin part leaching during interaction with water.

PF resin-treated plywood did not considerably lower growth rate of mould and blue stain compared to untreated plywood. However, we observed that mould and blue stain on the surface of PF resin treated plywood show a weaker colouration, resulting in a lower total colour change compared to untreated plywood (Figure 4 above). Surprisingly, treatment with 15 and 20% resin D solutions did not provide better surface protection for plywood against mould and blue stain. This was most likely due to the number of surface cracks that developed during the test period. All PF resin D treated specimens were more cracked compared to PF resin A, B and C treated specimens.

3.5. Resistance Against Soft-Rot in Unsterile Soil-Bed Test

Figure 5 below shows the mean ML_{wood} and standard deviation of the untreated plywood, birch and beech solid wood and PF-treated plywood specimens. For untreated birch plywood, beech solid wood and birch solid wood specimens, measurements of ML_{wood} were performed after 16, 20 and 24 weeks of incubation. Due to limitations in the number of treated test specimens, PF resin treated specimens were only removed after 16 and 24 weeks of incubation. For all control specimens, ML_{wood} increased almost as a linear function within the test period. Among the control specimens, the highest ML_{wood} range attained was for birch solid wood (26–38%), following with beech solid wood (22–34%), while the lowest was birch plywood (20–30%).

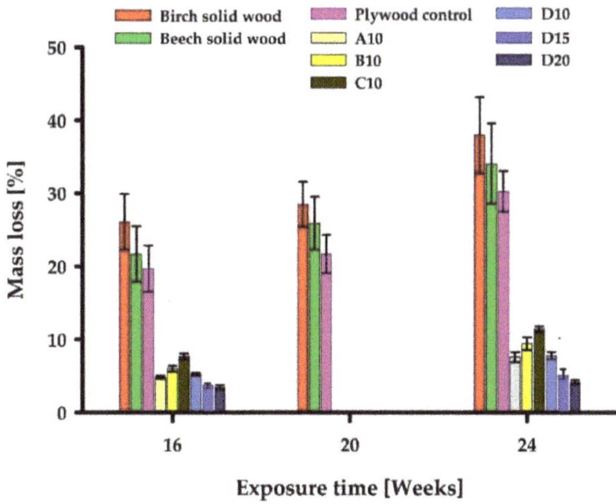

Figure 5. Oven-dry mass loss (ML_{wood}) [%] and standard deviation [%] of PF resin impregnated plywood material as well as commercially produced untreated plywood and solid wood after 16–24 weeks of incubation in unsterile soil.

The PF resin-treated plywood specimens showed a considerably reduced ML_{wood} compared to untreated references. The lowest ML_{wood} attained was for specimens treated with 10% solid content concentration of PF resins A and D (M_w = 292 and 703 g/mol), showing ML_{wood} of 4.8–7.6% and 5.3–7.8%, respectively. Specimens treated with resins B and C showed ML_{wood} of 6.0–9.4% and 7.6–11.4%, respectively. For laboratory prepared resin A, B, and C (M_w = 292,528 and 884 g/mol), a clear trend could be identified—use of higher molecular mass resins for veneer treatment caused higher ML_{wood} of plywood after exposure to unsterile soil. Specimens treated with commercially acquired resin D (M_w = 703 g/mol) at 10% solid content concentration did not fit into this trend since it showed lower ML_{wood} compared to PF resin B and C specimens treated at the same 10% solid content concentration. However, the assumption that an increase of resin loading in wood improves resistance against soft-rot was confirmed. The best results were shown by specimens treated with 15 and 20% solid content concentration solutions of PF resin D, with ML_{wood} of 3.7–5.1% and 3.4–4.1%, respectively.

x-Value towards Provisional Durability Rating

After 16 weeks of incubation, sufficient mean ML_{wood} (20%) of 10 reference specimens was reached for all reference specimens, exception for commercially sourced, untreated birch plywood. Subsequent incubation periods of 20 and 24 weeks also showed sufficient ML_{wood} for all untreated reference specimens. Median ML_{wood} was used to calculate an x value towards a provisional durability rating for the developed plywood material in accordance with CEN/TS 15083-2:2005 [28]. A provisional durability class (DC) rating was calculated after 16 and 24 weeks for all PF resin treated specimens against the various untreated reference control specimens (birch solid wood, beech solid wood and birch plywood). All PF treated plywood specimens could be classified as durable (DC 2) or moderately durable (DC 3). After treatment with 10% solid content concentration solutions, the best resistance to soft-rot was shown by resin A (M_w = 292 g/mol) treated specimens with DC 2 against birch solid wood after 16 and 24 weeks (lowest x value: Table 5). Plywood specimens treated with resins B and C (Mw = 528 and 884 g/mol) showed DC 3 against all controls at 16 and 24 weeks. Resin D (Mw = 703 g/mol) showed DC 2 relative to birch solid wood after 24 test weeks. Solid content concentration increases of resin D at 15 and 20% improved the resistance to soft-rotting fungi with specimens showing DC 2 against all reference materials after 16 and 24 weeks (Table 6).

Table 5. x values calculated according to CEN/TS 15083-2 [28] for the developed plywood material against various untreated reference wood materials for 16 and 24 weeks of incubation in unsterile soil.

Wood Material	Reference Material and Exposure Time [Weeks]					
	Birch Solid	Beech Solid	Untreated Plywood	Birch Solid	Beech Solid	Untreated Plywood
	16	16	16	24	24	24
A10	0.20	0.23	0.25	0.20	0.22	0.26
B10	0.25	0.28	0.31	0.24	0.26	0.31
C10	0.31	0.36	0.40	0.29	0.32	0.38
D10	0.22	0.25	0.27	0.20	0.22	0.26
D15	0.15	0.17	0.18	0.13	0.14	0.17
D20	0.14	0.15	0.17	0.11	0.12	0.14

Table 6. Durability class ratings according to CEN/TS 15083-2:2005 [28] after 16 and 24 weeks of incubation in unsterile soil.

Wood Material	Reference Material and Exposure Time [Weeks]					
	Birch Solid	Beech Solid	Untreated Plywood	Birch Solid	Beech Solid	Untreated Plywood
	16	16	16	24	24	24
A10	2	3	3	2	3	3
B10	3	3	3	3	3	3
C10	3	3	3	3	3	3
D10	3	3	3	2	3	3
D15	2	2	2	2	2	2
D20	2	2	2	2	2	2

4. Conclusions

In the Introduction we assumed that the weathering stability and resistance to biotic degradation agents of wood treated with low molecular weight PF resin would be improved, especially by using lower molecular weight PF resins, which in-turn increase resin loading in the wood cell wall.

Our results show that it is not possible to compare weathering processes that take place under artificial conditions with processes that take place under real, outdoor conditions. Even the combination of UV and water spraying under artificial conditions was not able to simulate similar material weathering processes as those achieved outdoors.

Birch veneers that were treated with laboratory synthesized PF resins A, B, and C (M_w = 292, 528 and 884 g/mol) prior to plywood production behaved better in both simulated and outdoor weathering conditions compared to plywood veneers treated with commercially sourced PF resin D. The assumption that increasing the concentration of resin in wood improves its weathering stability was not confirmed. Furthermore, it was also not possible to determine a clear trend of how mean M_w of a particular resin affects UV stability. Theoretically, resins with a higher molecular weight penetrate the wood cell wall less, so their concentration on the wood surface is potentially higher. Therefore, weathering stability of specimens treated with higher molecular weight resins could be expected to be better. However, our results do not support this theoretical consideration.

Resins A and D (M_w = 292 and 703 g/mol) seemed to be the most suitable for protection against soft-rot in the adapted unsterile soil bed test carried out in this study. For plywood treated with laboratory synthesized PF resins A, B, and C (Mw = 292, 528 and 884 g/mol), results suggest that the use of higher M_w resin increase ML wood after incubation in unsterile soil under various incubation periods. However, no clear correlation between M_w and ML_{wood} could be established when comparing all four resins used in this study-plywood veneers treated with resins A, B, C and D, at 10% solid content concentration. Specimens treated with resin D showed better resistance against soft-rot decay compared to resin B and C. It was confirmed that an increase of resin D loading in wood improves resistance against soft-rot.

Author Contributions: Conceptualization, J.G., V.B. and B.N.M.; Methodology, J.G., V.B. and B.N.M.; Formal Analysis, J.R. and H.M.; Investigation, J.G., V.B. and B.N.M.; Writing—Original Draft Preparation, J.G., V.B. and B.N.M.; Writing—Review and Editing, B.N.M., J.R. and H.M.; supervision, V.B. and J.R.; Project administration, J.G. All authors have read and agreed to the published version of the manuscript.

Funding: This research was funded by the Post-doctoral research project No.1.1.1.2/VIAA/1/16/210, agreement No. 1.1.1.2/16/I/001.

Acknowledgments: Special thanks to Prefere Resins Germany GmbH Erkner and personally Elke Fliedner, Klaus Dück and Christopher Knie for providing PF resins and for help with the GPC analysis.

Conflicts of Interest: The authors declare no conflict of interest.

References

1. Lamb, F.M. Splits and Cracks in Wood. In Proceedings of the 43rd Annual Meeting of Western Dry Kiln Association, Reno, NV, USA, 13–15 May 1992; pp. 16–24.
2. Evans, P.D. Weathering and photoprotection of wood. In *Development of Commercial Wood Preservatives: Efficacy, Environmental, and Health Issues*; American Chemical Society; ACS Symposium Series; Schultz, T.P., Militz, H., Freeman, M.H., Goodell, B., Nicholas, D.D., Eds.; Oxford University Press: Washington, DC, USA, 2008; pp. 69–119. ISBN 978-0-8412-3951-7.
3. Altgen, M.; Adamopoulos, S.; Militz, H. Wood Defects during Industrial-Scale Production of Thermally Modified Norway Spruce and Scots Pine. *Wood Mater. Sci. Eng.* **2015**, *12*, 14–23. [CrossRef]
4. *EN 350:2016 Durability of Wood and Wood-Based Products—Testing and Classification of the Durability to Biological Agents of Wood and Wood-Based Materials*; European Committee for Standardization (CEN): Brussels, Belgium, 2016.
5. Helin, T.; Sokka, L.; Soimakallio, S.; Pingoud, K.; Pajula, T. Approaches for Inclusion of Forest Carbon Cycle in Life Cycle Assessment—A Review. *GCB Bioenergy* **2013**, *5*, 475–486. [CrossRef]
6. Hill, C.A.S. *Wood Modification: Chemical, Thermal and Other Processes*; John Wiley & Sons, Ltd.: Chichester, UK, 2006; ISBN 978-0-470-02174-3.
7. Xie, Y.; Krause, A.; Militz, H.; Turkulin, H.; Richter, K.; Mai, C. Effect of Treatments with 1,3-Dimethylol-4,5-Dihydroxy-Ethyleneurea (DMDHEU) on the Tensile Properties of Wood. *Holzforschung* **2007**, *61*, 43–50. [CrossRef]
8. Rowell, R.; Bongers, F. Coating Acetylated Wood. *Coatings* **2015**, *10*, 792–801. [CrossRef]
9. Evans, P.D.; Wallis, A.F.A.; Owen, N.L. Weathering of Chemically Modified Wood Surfaces. *Wood Sci. Technol.* **2000**, *34*, 151–165. [CrossRef]
10. Xiao, Z.; Xie, Y.; Adamopoulos, S.; Mai, C. Effects of Chemical Modification with Glutaraldehyde on the Weathering Performance of Scots Pine Sapwood. *Wood Sci. Technol.* **2012**, *46*, 749–767. [CrossRef]
11. Feist, W.C.; Sell, J. Weathering Behavior of Dimensionally Stabilized Wood Treated by Heating under Pressure of Nitrogen Gas. *Wood Fiber Sci.* **1987**, *19*, 13.
12. Furuno, T.; Imamura, Y.; Kajita, H. The Modification of Wood by Treatment with Low Molecular Weight Phenol-Formaldehyde Resin: A Properties Enhancement with Neutralized Phenolic-Resin and Resin Penetration into Wood Cell Walls. *Wood Sci. Technol.* **2004**, *37*, 349–361. [CrossRef]
13. Tarkow, H.; Southerland, C.F.; Seborg, R.M. *Surface Characteristics of Wood as They Affect Durability of Finishes Part 1. Surface Stabilizazion*; FPL 57; U.S. Department of Agriculture, U.S. Forest Service: Madison, WI, USA, 1966; pp. 3–22.
14. Morawetz, H. Phenolic Antioxidants for Paraffinic Materials. *Ind. Eng. Chem.* **1949**, *41*, 1442–1447. [CrossRef]
15. Kielmann, B.C.; Mai, C. Application and Artificial Weathering Performance of Translucent Coatings on Resin-Treated and Dye-Stained Beech-Wood. *Prog. Org. Coat.* **2016**, *95*, 54–63. [CrossRef]
16. Kielmann, B.C.; Mai, C. Natural Weathering Performance and the Effect of Light Stabilizers in Water-Based Coating Formulations on Resin-Modified and Dye-Stained Beech-Wood. *J. Coat. Technol. Res.* **2016**, *13*, 1065–1074. [CrossRef]
17. Evans, P.D.; Gibson, S.K.; Cullis, I.; Liu, C.; Sèbe, G. Photostabilization of Wood Using Low Molecular Weight Phenol Formaldehyde Resin and Hindered Amine Light Stabilizer. *Polym. Degrad. Stab.* **2013**, *98*, 158–168. [CrossRef]

18. Kielmann, B.C.; Butter, K.; Mai, C. Modification of Wood with Formulations of Phenolic Resin and Iron-Tannin-Complexes to Improve Material Properties and Expand Colour Variety. *Eur. J. Wood Prod.* **2018**, *76*, 259–267. [CrossRef]

19. Van Acker, J.; Palanti, S. 5.3 Durability. In *Performance of Bio-Based Building Materials*; Woodhead Publishing Series in Civil and Structural Engineering; Europäische Zusammenarbeit auf dem Gebiet der Wissenschaftlichen und Technischen Forschung; Brischke, C., Jones, D., Eds.; WP—Woodhead Publishing, an Imprint of Elsevier: Cambridge, UK, 2017; pp. 257–277. ISBN 978-0-08-100992-5.

20. Zabel, R.A.; Morrell, J.J. 4 Factors Affecting the Growth and Survival of Fungi in Wood. In *Wood Microbiology: Decay and Its Prevention*; Elsevier Science: Saint Louis, MI, USA, 2014; pp. 90–115. ISBN 978-0-323-13946-5.

21. Edlund, M.-L.; Nilsson, T. Testing the Durability of Wood. *Mat. Struct.* **1998**, *31*, 641–647. [CrossRef]

22. Brischke, C.; Olberding, S.; Meyer, L.; Bornemann, T.; Welzbacher, C.R. Intrasite Variability of Fungal Decay on Wood Exposed in Ground Contact. *Int. Wood Prod. J.* **2013**, *4*, 37–45. [CrossRef]

23. Torres-Andrade, P.; Morrell, J.J.; Cappellazzi, J.; Stone, J.K. Culture-Based Identification to Examine Spatiotemporal Patterns of Fungal Communities Colonizing Wood in Ground Contact. *Mycologia* **2019**, *111*, 703–718. [CrossRef] [PubMed]

24. Nilsson, T.; Daniel, G. *Decay Types Observed in Small Stakes of Pine and Alstonia Scholaris Inserted in Different Types of Unsterile Soil, IRG/WP/1443*; IRG Secretariat: Stockholm, Sweden, 1990; p. 10.

25. Mieß, S. Einfluß Des Wasserhaushaltes Auf Abbau Und Fäuletypen von Holz in Terrestrichen Mikrokosmen (Diplomarbeit). Bachelor's Thesis, Universität Hamburg Ordinariat für Holzbiologie, Hamburg, Germany, 1997.

26. *EN 252:2015. Field Test Methods for Determining the Relative Protective Effectiveness of Wood Preservatives in Ground Contact*; European Committee for Standardization (CEN): Brussels, Belgium, 2014.

27. *CEN/TS 15083-1:2005 Durability of Wood and Wood-Based Products—Determination of the Natural Durability of Solid Wood against Wood-Destroying Fungi, Test Methods—Part 1: Basidiomycetes*; European Committee for Standardization (CEN): Brussels, Belgium, 2005.

28. *CEN/TS 15083-2:2005 Durability of Wood and Wood-Based Products—Determination of the Natural Durability of Solid Wood against Wood-Destroying Fungi, Test Methods—Part 2: Soft Rotting Micro-Fungi*; European Committee for Standardization (CEN): Brussels, Belgium, 2005.

29. *DIN EN ISO 3251:2019 Paints, Varnishes and Plastics; Determination of Non-Volatile-Matter Content*; European Committee for Standardization (CEN): Brussels, Belgium, 2019; p. 12.

30. *DIN EN ISO 9397:1997 Plastics—Phenolic Resins—Determination of Free-Formaldehyde Content—Hydroxylamine Hydrochloride Method*; European Committee for Standardization (CEN): Brussels, Belgium, 1997.

31. Grinins, J.; Biziks, V.; Irbe, I.; Rizhikovs, J. Water Related Properties of Birch Wood Modified with Phenol-Formaldehyde (PF) Resins. *Key Eng. Mater.* **2018**, *800*, 246–250. [CrossRef]

32. Grinins, J.; Irbe, I.; Biziks, V.; Rizikovs, J.; Bicke, S.; Militz, H. Investigation of Birch Wood Impregnation with Phenol-Formaldehyde (PF) Resins. In Proceedings of the 9th European Conference on Wood Modification (ECWM9), Arnhem, The Netherlands, 17–18 September 2018; p. 6.

33. *DIN EN 152-1:2011 Test Methods for Wood Preservatives; Laboratory Method for Determining the Protective Effectiveness of a Preservative Treatment against Blue Stain in Service, Part 1: Brushing Procedure*; European Committee for Standardization (CEN): Brussels, Belgium, 2011.

34. *ISO 11268-2 Soil Quality—Effects of Pollutants on Earthworms—Part 2: Determination of Effects on Reproduction of Eisenia Fetida/Eisenia Andrei*; International Organisation for Standardization (ISO): Geneva, Switzerland, 2012.

35. *EN 335 Durability of Wood and Wood-Based Products—Use Classes: Definition, Application to Solid Wood and Wood-Based Products*; European Committee for Standardization (CEN): Brussels, Belgium, 2013.

36. *EN 460:1994 Durability of Wood and Wood-Based Products—Natural Durability of Solid Wood—Guide to the Durability Requirements for Wood to Be Used in Hazard Classes*; European Committee for Standardization (CEN): Brussels, Belgium, 1994.

37. *PrEN 460:2019 Durability of Wood and Wood-Based Products—Natural Durability of Solid Wood—Guide to the Durability Requirements for Wood to Be Used in Hazard Classes*; European Committee for Standardization (CEN): Brussels, Belgium, 2019.

 *polymers*

Article

An Additive Manufacturing Method Using Large-Scale Wood Inspired by Laminated Object Manufacturing and Plywood Technology

Yubo Tao [1], Qing Yin [1] and Peng Li [1,2,*]

[1] State Key Laboratory of Biobased Material and Green Papermaking, Qilu University of Technology, Shadong Academy of Sciences, Jinan 250353, China; taoyubo@qlu.edu.cn (Y.T.); qluyinqing@163.com (Q.Y.)
[2] College of Material Science and Engineering, Northeast Forestry University, Harbin 150040, China
* Correspondence: lipeng@qlu.edu.cn

Abstract: Wood-based materials in current additive manufacturing (AM) feedstocks are primarily restricted to the micron scale. Utilizing large-scale wood in existing AM techniques remains a challenge. This paper proposes an AM method—laser-cut veneer lamination (LcVL)—for wood-based product fabrication. Inspired by laminated object manufacturing (LOM) and plywood technology, LcVL bonds wood veneers in a layer-upon-layer manner. As demonstrated by printed samples, LcVL was able to retain the advantageous qualities of AM, specifically, the ability to manufacture products with complex geometries which would otherwise be impossible using subtractive manufacturing techniques. Furthermore, LcVL-product structures designed through adjusting internal voids and wood-texture directionality could serve as material templates or matrices for functional wood-based materials. Numerical analyses established relations between the processing resolution of LcVL and proportional veneer thickness (layer height). LcVL could serve as a basis for the further development of large-scale wood usage in AM.

Keywords: veneer; laser-cut; additive manufacturing; wood composite

Citation: Tao, Y.; Yin, Q.; Li, P. An Additive Manufacturing Method Using Large-Scale Wood Inspired by Laminated Object Manufacturing and Plywood Technology. *Polymers* **2021**, *13*, 144. https://doi.org/10.3390/polym13010144

Received: 19 December 2020
Accepted: 28 December 2020
Published: 31 December 2020

Publisher's Note: MDPI stays neutral with regard to jurisdictional claims in published maps and institutional affiliations.

1. Introduction

The controlled process of material removal is a definitive trait of subtractive manufacturing technologies (SM). Traditional wood-processing techniques such as sawing, milling, turning, carving, and grinding, as well as relatively modern techniques such as CNC (Computer Numerical Control), are all categorized as SM [1]. As shown in Figure 1, portions of the raw material are methodically removed until the intended shape is achieved. By contrast, additive manufacturing (AM), as shown in Figure 1, often referred to as 3D printing, is a process of joining materials, typically in a layer-upon-layer manner, in accordance with three-dimensional (3D) model data [2]. Fabrication using AM begins with a 3D model of the desired product, such as the model shown in Figure 2a,b. Subsequently, 3D printing software will slice the model into horizontal cross-sectional layers, as shown in Figure 2c. Ultimately, the model is fabricated by stacking layers, an example of which is shown in Figure 2d.

Variations of AM are differentiated by their respective layer-fabrication techniques, including stereolithography apparatus (SLA), fused deposition modeling (FDM), laminated object manufacturing (LOM), selective laser sintering (SLS), and direct energy deposition (DED) [3]. Notably, AM is especially advantageous compared to SM when manufacturing products with exceptional geometric complexity. Currently, AM technologies have extended to areas in the aerospace, automotive, medical, architecture, and fashion industries [4]. The continuously increasing demand for renewable and sustainable products sourced from petroleum-free and carbon-neutral origins has driven the development of novel materials for AM methods in recent years.

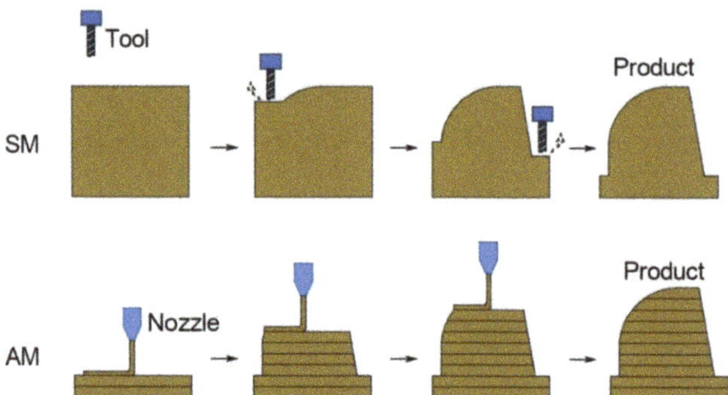

Figure 1. Subtractive manufacturing (SM) and additive manufacturing (AM).

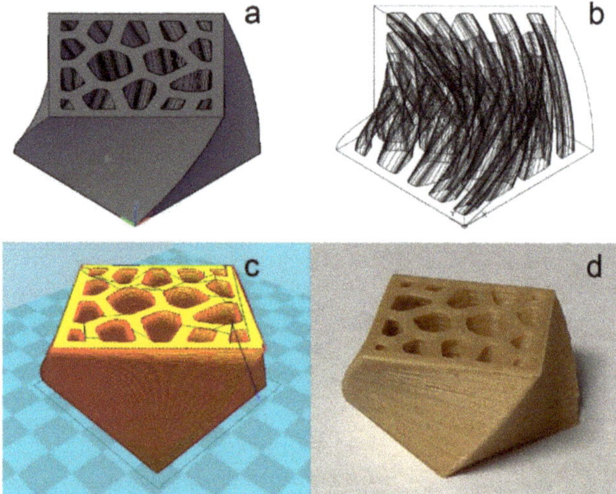

Figure 2. An illustration of the typical AM process. (**a**) 3D model of the desired product; (**b**) "wireframe" display of the desired product; (**c**) model after slicing into layers by the CURA software; (**d**) desired product constructed using fused deposition modeling (FDM) 3D printing.

Wood derivatives, such as wood flour and sawdust, as well as the components of wood, i.e., cellulose and lignin, are naturally abundant, biodegradable, biocompatible, and chemically modifiable materials that have shown promising potential for AM [5,6]. Existing research has shown that the practicability of incorporating wood-based materials in AM is largely dependent on the respective AM technique [7–13]. At present, layer fabrication techniques using wood-based materials may be divided into two general categories: extrusion-deposition and granular bonding. Extrusion-deposition fabrication primarily employs wood-plastic composite filaments that could be used in FDM [7,8]. In addition, studies have also shown that it is possible to extrude and deposit a slurry mixture of sawdust and adhesive directly to achieve similar AM results [9–11]. Likewise, granular bonding comprises two distinct variants. One type involves melting powdered mixtures of thermoplastic polymers and wood-based materials with high-intensity lasers [12], a technique utilized by SLS, whereas the other relies on the solidification reaction of a wood-based bulk material, as inorganic binders blend upon contact with water [13].

LOM is one of the first commercially available AM techniques, in which sheets of material, including metal, plastic, and paper, etc., are cut, often with lasers or mechanical cutters, to precisely resemble the shape of the cross-sections of the desired product. Successive layers are bonded layer upon layer until the object is completed [14,15]. Nevertheless, wood-based product fabrication with the aforementioned AM techniques is primarily dominated by micron scale powder and fiber materials. Current preparation methodologies not only increase the overall processing difficulty of wood-based materials, but also create drastic discrepancies, in both appearance and mechanical properties, compared with the original wood.

The utilization of large-scale wood materials in AM has rarely been explored. Existing studies have investigated the application of one-dimensional wood-based materials, such as sticks and strips, in AM. For example, one study involved dispensing chopsticks coated in wood adhesive from a projection mapping-guided handheld stick dispenser to construct architectural structures [16]. Another study fabricated high-resolution timber structures with continuous willow withe-based solid wood filaments, a robotic fiber placement process, and topology optimization [17].

This paper proposes ideas for an alternative AM method for wood-based product fabrication that would be able to utilize large scale wood-based materials, such as wood veneer (a two-dimensional surface), by combining plywood technology with the basis behind LOM [18,19]. In addition to granular and strip-like, wood-based AM materials, the proposed method could enable the use of plate-like wood materials in AM. Furthermore, this study is characterized by the use of simple processing techniques, such as cutting and gluing, and AM characteristics to manufacture wood products with complex shapes and internal structures without advanced subtractive techniques, such as robotic CNC engraving. Moreover, its AM capabilities could be used for creating designable templates and material matrices for functional wood-based materials, such as sound absorbers and composites. Inspired by LOM, this process can be named laser-cut veneer lamination (LcVL), in which sheets of laser-cut veneer form cross-sectional layers that are bonded layer upon layer to form wood products with complex geometries and internal voids.

2. Materials and Methods

An LcVL-printed product was fabricated based on the design shown in Figure 2 to demonstrate the capabilities of the proposed AM method.

2.1. Modeling

The procedures used in the construction of a 3D model of the sample were as follows: as depicted in Figure 3a, a 50 mm × 50 mm square was created on the XOY plane (AutoCAD, student version 2019, San Rafael, CA, USA). The interior of this square was then partitioned into 16 Voronoi cells. An extrusion of 1.5 mm was applied to the surface along the Z-axis to create a layer model for the sample, as shown in Figure 3d. A total of 20 duplicates of the layer model were stacked along the Z-axis, as illustrated in Figure 3e. Lastly, all layers underwent rotation with the angle of rotation increment by 2.25° with each passing layer, as shown in Figure 3f.

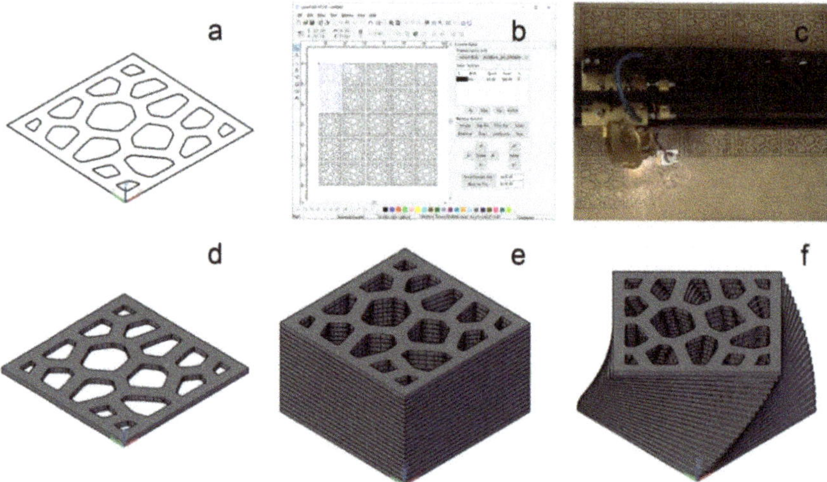

Figure 3. Model design methodology. (**a**) Outline of a Voronoi cellular-patterned cross-sectional layer; (**b**) Setting laser-processing parameters in the LaserCAD software, such as laser power, moving speed, etc.; (**c**) Laser-cut wood; (**d**) A layer slice after 1.5 mm of extrusion; (**e**) Stacking of layers along the Z-axis to create a layered model; (**f**) Layers are rotated to produce the model of the desired product.

2.2. Processing

Poplar (*Aspen*) veneer with a nominal thickness of 1.5 mm and 8% moisture content was adopted in this work. The design shown in Figure 3a was fed to the LaserCAD software (Shenzhen Qiancheng Co., Ltd., Shenzhen, China) for setting laser processing parameters such as path, power, and speed. As shown in Figure 3c, a laser-carving machine (Model 4060, Huitian Laser Instrument Co., Ltd., Jinan, China) was used to cut veneers following the path and parameters set in Figure 3b to create each layer of the desired product.

The top of each layer was coated with polyvinyl acetate (PVA) adhesive (Pattex 710, Pattex Co., Ltd., Shanghai, China) before being stacked to form a mat in accordance with the model design. A mold of the model contour could be used to guarantee layer placement precision. After 2 min of deposition, the mat was pressed for 5 min under 10 N using a small cold presser (lab-made) to complete the bonding process. ifferent adhesives could be used with adjusted pressing parameters.

3. Results and Discussion

3.1. The LcVL Product

As shown in Figure 4, the LcVL product was fabricated by stacking and bonding wood veneers in a layer-upon-layer manner. The product demonstrated that the LcVL procedure was able to take advantage of the qualities of additive manufacturing, specifically, the ability to manufacture complex geometries, such as internal voids, that are nigh impossible to accomplish using SM techniques, such as CNC. However, since LcVL is based on LOM characteristics, although the overall product formation is additive in nature, the production of each layer via cutting is a subtractive process. These subtractive drawbacks should be marginal in comparison to the technical simplicity of the LcVL process. Residual materials could be repurposed as raw materials for 3D printing. For example, leftover veneers could be used to produce wood powder for wood/polylactic acid filaments.

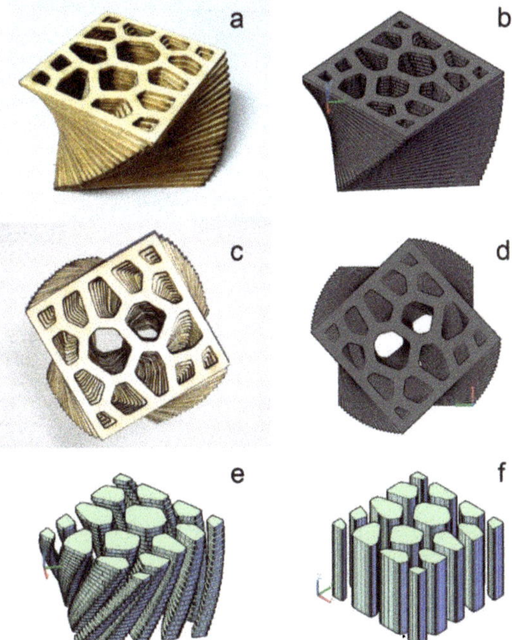

Figure 4. Comparison between the laser-cut veneer lamination (LcVL) product and its 3D model. (**a**) Orthographic view of the LcVL product; (**b**) Orthographic view of the 3D model of the product; (**c**) Top view of the LcVL product; (**d**) Top view of the 3D model of the product; (**e**) Tubular voids present in the 3D model of the product postrotation; (**f**) Tubular voids present in the 3D model of the product prerotation.

Furthermore, the surface area of tubular voids present in Figure 4e is 1.27 times the surface area of SM possible tubular voids in Figure 4f. The increased surface area in products with intricate geometrical structures, such as the product presented in Figure 4, could prove beneficial for the development of special-purpose, wood-based products. For example, the spacious tubular voids of complex LcVL-printed structures contain larger void surfaces and enable greater convenience for architecting desired tortuosity, which could improve sound absorption compared to standard SM possible structures [20]. Overall, as demonstrated by the printed product in Figure 4, LcVL was able to properly realize the 3D model of the desired product to a satisfactory degree. However, the LcVL method is not ideal for fabricating products with high angle overhangs without additional external support to ensure uniform pressure on each layer.

Notably, comparing the printed models present in Figure 4a (LcVL) and Figure 2a,d (FDM) revealed visible distinctions in processing resolution. As will be discussed in detail in the following section, the fabricating resolution of LcVL-printed products is primarily dependent on the layer parameters.

3.2. Effects of Layer Parameters on Processing Resolution

As depicted in Figure 5a, LcVL is unable to replicate the modeling curve line (MCL) with perfect precision. The resulting step-like contour along the Z-axis comprises a theoretical manufacturing error (TME) between the 3D model and the fabricated product. Using the region circled in green in Figure 5a as an example, the relation between layer (veneer) height and TME could be described as follows:

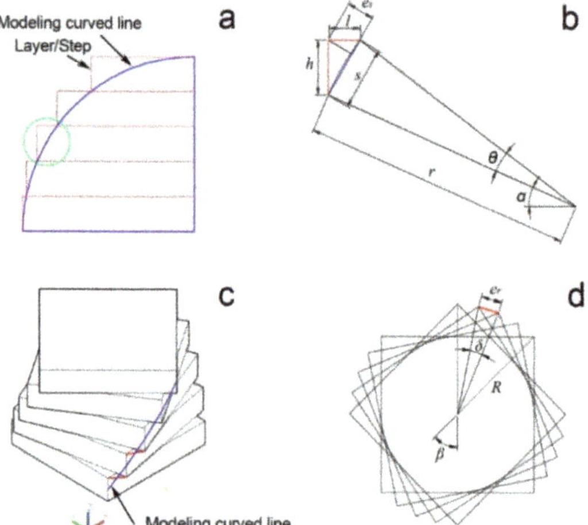

Figure 5. (**a**) Theoretical manufacturing error between contours of the modeling curve line and LcVL layer stacking; (**b**) Calculation parameters for the theoretical manufacturing error from LcVL layer stacking; (**c**) Theoretical manufacturing error between contours of the modeling curve line and postrotation LcVL layer stacking; (**d**) Calculation parameters for the theoretical manufacturing error from postrotation layer stacking.

As shown in Figure 5b, when s is the arc length of a MCL, r the radius of the MCL, θ the central angle of the MCL, α the angle of the MCL to horizontal, h the layer height, and e_s the TME from layer height, then

$$s = 2r \sin \frac{\theta}{2} \tag{1}$$

$$l = \sqrt{s^2 - h^2} \tag{2}$$

$$\frac{1}{2} s \cdot e = \frac{1}{2} h \cdot l \tag{3}$$

$$e_s = \frac{h \cdot \sqrt{s^2 - h^2}}{s} \tag{4}$$

The relation between layer height and TME from each step/layer of a quarter circle MCL with radius 1 was calculated and plotted in Figure 6a. The quarter circle was divided into five and ten layers to obtain proportional layer heights of 0.2 and 0.1, respectively. As can be seen in Figure 6a, the proportional layer height of 0.2 consistently exhibited greater TME compared to the smaller layer height of 0.1. Therefore, TME is positively associated with layer height.

In addition to the TME caused by layer height, the fabrication accuracy of the LcVL product in Figure 4 suffered further TME from layer rotation. As shown in Figure 5c, the apparent discrepancy between the MCL (highlighted in blue) and the printed product contributed to additional TME (highlighted in red). The TME from layer rotation could be described as follows:

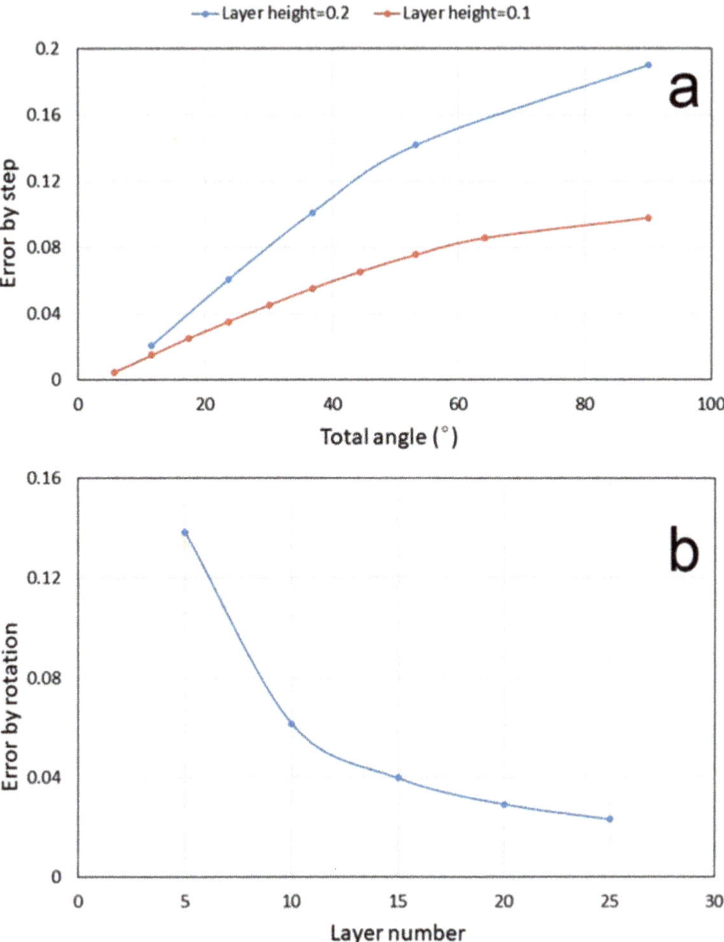

Figure 6. (**a**) Relation between layer height and theoretical manufacturing error (TME) from layer height, where total angle is the sum of α and θ; (**b**) Relation between layer number and TME from layer rotation.

As shown in Figure 5d, when R is the radius of rotation, n is the number of layers (layer number), β is the angle of rotation between the top and bottom layers, δ is the angle of rotation between successive layers, and e_r is the TME from layer rotation, then

$$\delta = \frac{\beta}{n-1} \tag{5}$$

$$e_r = 2R \sin \frac{\delta}{2} \tag{6}$$

$$e_r = 2R \sin \frac{\beta}{2(n-1)} \tag{7}$$

The relation between layer numbers (5–25) and TME from layer rotation for a height-1 hypothetical model with 45° of rotation between the top and bottom layers was calculated and plotted in Figure 6b. As can be seen in Figure 6b, the corresponding TME from layer rotation underwent reduction with larger layer numbers. Therefore, for the same product height, larger layer (veneer) numbers could result in decreased TME not only from layer rotation, but also from layer height as a result of the smaller layer height. Notably, for the product presented in this study (Figure 4), calculations showed that a 100% increase in layer number could increase the bonding area by 225%, which could increase production costs. The lower the layer height, the smaller the veneer thickness, which also increases the difficulty of veneer manufacturing. Notably, although the sample created for this study was a small object in the centimeter scale, the core characteristics of the LcVL method could be scaled up to manufacture structures in the meter scale, in theory. Naturally, corresponding parameters, such as the product height, layer height, and layer number should be adjusted accordingly to optimize the TME.

3.3. Wood Texture Direction and LcVL-Product Structure

The structural directionality of LcVL products could be designed through wood texture directions. The sample presented in Figure 4 was created by stacking identical layers with each layer rotated by 2.25°. As shown in Figure 7a, a pair of identically-cut layers share the same wood texture direction. Thus, since all layers are 2.25° offset from their adjacent layers, the wood texture directions of all layers are 2.25° apart in this sample. However, as shown in Figure 7b, if layer 2 was cut with a counterclockwise 2.25° rotation from layer 1, then the wood texture direction of layer 2 would be 2.25° clockwise from layer 1. Thus, if such layer pairs were laminated together with a 2.25° counterclockwise layer-to-layer increment, the resulting product would have consistent wood texture direction. Alternatively, if layer 2 was cut with a 90° counterclockwise plus 2.25° counterclockwise rotation, as shown in Figure 7c, then the wood texture directions of layers 1 and 2 would be orthogonal in a product with 2.25° counterclockwise-rotated layers. The directionality of such a product could be analyzed with the orthogonal principle of plywood technology. The designability of LcVL-product structures is essential for creating material templates and matrices for composites of varying properties with LcVL.

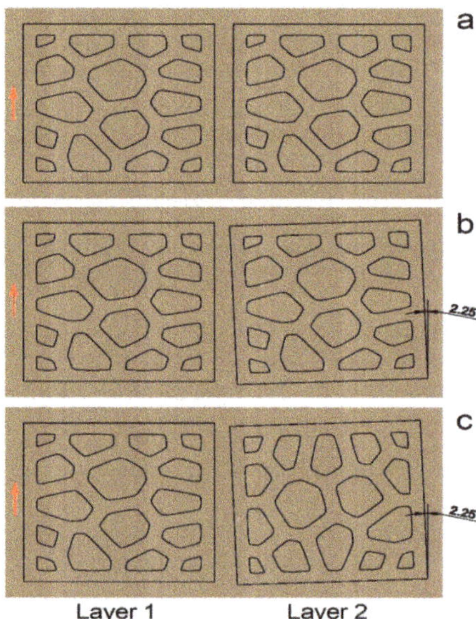

Figure 7. Wood-texture direction (indicated by red arrows) and distinct layer-cutting solutions (**a**) Cutting solution with identically cut layer 1 and layer 2. The resulting wood texture directions of the structure are 2.25° between each layer. (**b**) Cutting solution that produces a product with a consistent wood texture direction. (**c**) Cutting solution that produces a product with orthogonal wood texture directions between layers.

4. Conclusions

LcVL is a relatively simple procedure for constructing customized and geometrically complex wood products which would otherwise be impossible, difficult, and/or costly using SM. Aside from raw material costs, costs of material waste and time consumption are optimizable factors of efficiency. In addition, the LcVL produce should be fairly adoptable, as its core technologies, laser-cutting and plywood, are already widely used in the wood industry.

From the above findings, the following conclusions could be made:

(1) As a combination of plywood technology and LOM, the LcVL method is a viable AM method which is capable of producing wood-based products with complex geometries and internal voids using large scale wood-based materials, specifically wood veneer.

(2) LcVL products have designable structures (complex internal voids and wood texture directions). Their designability could be used for creating material matrices and templates for functional wood-based materials, such as sound absorbers and composites.

(3) The LcVL method encountered more theoretical manufacturing errors compared to other AM techniques due to its use of larger scale raw materials with larger layer heights. Nonetheless, the LcVL method may be used for large scale wood materials with sufficient layer thickness and number.

(4) LcVL products could benefit greatly from postprocessing, such as surface finishing, for theoretical manufacturing-error reduction. In comparison with other AM techniques, larger amounts of wood and less adhesives are involved during the fabrication. The LcVL method could serve as a basis for the further development of veneer usage in AM.

Author Contributions: Conceptualization, Y.T. and P.L.; methodology, P.L.; investigation, Y.T. and Q.Y.; writing—original draft preparation, Y.T.; writing—review and editing, Y.T. and P.L.; All authors have read and agreed to the published version of the manuscript.

Funding: This project was supported by the PROGRAM FOR NEW CENTURY EXCELLENT TALENTS IN UNIVERSITY OF CHINA, grant number NCET-13-0711 and the START-UP FUNDING FROM QILU UNIVERSITY OF TECHNOLOGY, SHANDONG ACADEMY OF SCIENCES.

Institutional Review Board Statement: Not applicable.

Informed Consent Statement: Not applicable.

Data Availability Statement: Data sharing not applicable.

Acknowledgments: The authors would like to thank Zelong Li (University of British Columbia) for his help.

Conflicts of Interest: The authors declare no conflict of interest.

References

1. Wimmer, R.; Steyrer, B.; Woess, J.; Koddenberg, T.; Mundigler, N. 3D printing and wood. *Pro Ligno* **2015**, *11*, 144–149. Available online: https://publik.tuwien.ac.at/files/PubDat_243647.pdf (accessed on 5 November 2015).
2. *ASTM F2792-12a. Standard Terminology for Additive Manufacturing Technologies*; ASTM International: West Conshohocken, PA, USA, 2012. [CrossRef]
3. Ngo, T.D.; Kashani, A.; Imbalzano, G.; Nguyen, K.T.Q.; Hui, D. Additive manufacturing (3D printing): A review of materials, methods, applications and challenges. *Compos. B Eng.* **2018**, *143*, 172–196. [CrossRef]
4. Gardan, J. Additive manufacturing technologies: State of the art and trends. *Int. J. Prod. Res.* **2016**, *54*, 3118–3132. [CrossRef]
5. Douglas, G.; Wang, L.; Wang, J. Additive Manufacturing of Wood-Based Materials for Composite Applications. In Proceedings of the SPE Automotive Composites Conference & Exhibition, Novi, MI, USA, 4–6 September 2019.
6. Li, T.; Aspler, J.; Kingsland, A.; Cormier, L.M.; Zou, X. 3D printing–a review of technologies, markets, and opportunities for the forest industry. *J. Sci. Technol. Prod. Process.* **2016**, *5*, 30–31.
7. Tao, Y.; Wang, H.; Li, Z.; Li, P.; Shi, S.Q. Development and application of wood flour-filled polylactic acid composite filament for 3d printing. *Materials* **2017**, *10*, 339. [CrossRef] [PubMed]
8. Zhao, X.; Tekinalp, H.; Meng, X.; Ker, D.; Benson, B.; Pu, Y.; Ragauskas, A.J.; Wang, Y.; Li, K.; Webb, E.; et al. Poplar as biofiber reinforcement in composites for large-scale 3D printing. *ACS Applied. Bio. Mater.* **2019**, *2*, 4557–4570. [CrossRef]

9.   Gardan, J.; Nguyen, D.C.; Roucoules, L.; Montay, G. Characterization of wood filament in additive deposition to study the mechanical behavior of reconstituted wood products. *J. Eng. Fiber Fabr.* **2016**, *11*, 56–63. [CrossRef]

10.   Kariz, M.; Sernek, M.; Kuzman, M.K. Use of wood powder and adhesive as a mixture for 3D printing. *Eur. J. Wood Wood Prod.* **2016**, *74*, 123–126. [CrossRef]

11.   Rosenthal, M.; Henneberger, C.; Gutkes, A.; Bues, C. Liquid Deposition Modeling: A promising approach for 3D printing of wood. *Eur. J. Wood Wood Prod.* **2018**, *76*, 797–799. [CrossRef]

12.   Zeng, W.; Guo, Y.; Jiang, K.; Yu, Z.; Liu, Y.; Shen, Y.; Deng, J.; Wang, P. Laser intensity effect on mechanical properties of wood–plastic composite parts fabricated by selective laser sintering. *J. Thermoplast. Compos.* **2013**, *26*, 125–136. [CrossRef]

13.   Henke, K.; Treml, S. Wood based bulk material in 3D printing processes for applications in construction. *Eur. J. Wood Wood Prod.* **2013**, *71*, 139–141. [CrossRef]

14.   Feygin, M.; Pak, S.S. Laminated Object Manufacturing Apparatus and Method. U.S. Patent 5,876,550, 2 March 1999.

15.   Wimpenny, D.I.; Bryden, B.; Pashby, I. Rapid laminated tooling. *J. Mater. Process. Technol.* **2003**, *138*, 214–218. [CrossRef]

16.   Yoshida, H.; Igarashi, T.; Obuchi, Y.; Takami, Y.; Sato, J.; Araki, M.; Miki, M.; Nagata, K.; Sakai, K.; Igarashi, S. Architecture-scale human-assisted additive manufacturing. *ACM Trans. Graph.* **2015**, *34*, 88. [CrossRef]

17.   Dawod, M.; Deetman, A.; Akbar, Z.; Heise, J.; Böhm, S.; Klussmann, H.; Eversmann, P. *Continuous Timber Fibre Placement. Impact: Design with All Senses*; DMSB 2019; Springer: Cham, Switzerland, 2020. [CrossRef]

18.   Eltawahni, H.A.; Rossini, N.S.; Dassisti, M.; Alrashed, K.; Aldaham, T.A.; Benyounis, K.Y.; Olabi, A.G. Evaluation and optimization of laser cutting parameters for plywood materials. *Opt. Laser Eng.* **2013**, *9*, 1029–1043. [CrossRef]

19.   Kubovský, I.; Kačík, F. Colour and chemical changes of the lime wood surface due to CO_2 laser thermal modification. *Appl. Surf. Sci.* **2014**, *321*, 261–267. [CrossRef]

20.   Errico, F.; Ichchou, M.; De Rosa, S.; Franco, F.; Bareille, O. Investigations about periodic design for broadband increased sound transmission loss of sandwich panels using 3D-printed models. *Mech. Syst. Signal.* **2020**, *136*, 106432. [CrossRef]

 *polymers*

Article

Investigation of 3D-Moldability of Flax Fiber Reinforced Beech Plywood

Johannes Jorda [1,2]**, Günther Kain** [1,*]**, Marius-Catalin Barbu** [1,3]**, Matthias Haupt** [1] **and Ľuboš Krišťák** [4]

[1]   Forest Products Technology and Timber Construction Department, Salzburg University of Applied Sciences, Markt 136a, 5431 Kuchl, Austria; jjorda.lba@fh-salzburg.ac.at (J.J.); marius.barbu@fh-salzburg.ac.at (M.-C.B.); mhaupt.htw-m2016@fh-salzburg.ac.at (M.H.)
[2]   Department of Wood Science and Technology, Mendel University, Zemědělská 3, 61300 Brno, Czech Republic
[3]   Faculty for Furniture Design and Wood Engineering, Transilvania University of Brasov, B-dul. Eroilor nr. 29, 500036 Brasov, Romania
[4]   Faculty of Wood Sciences and Technology, Technical University in Zvolen, T. G. Masaryka 24, SK-960 01 Zvolen, Slovakia; kristak@tuzvo.sk
*   Correspondence: gkain.lba@fh-salzburg.ac.at; Tel.: +43-699-819-764-42

Received: 30 October 2020; Accepted: 26 November 2020; Published: 29 November 2020

Abstract: The current work deals with three dimensionally molded plywood formed parts. These are prepared in two different geometries using cut-outs and relief cuts in the areas of the highest deformation. Moreover, the effect of flax fiber reinforcement on the occurrence and position of cracks, delamination, maximum load capacity, and on the modulus of elasticity is studied. The results show that designs with cut-outs are to be preferred when molding complex geometries and that flax fiber reinforcement is a promising way of increasing load capacity and stiffness of plywood formed parts by respectively 76 and 38% on average.

Keywords: plywood; veneer 3D moldability; natural fiber reinforcement

---

1. Introduction

The world is not flat-driven by the rising consumer awareness for the ecological product footprint, designers are set to bid and overcome the limits for material applications. Composite materials are developed to surpass single inferior material properties in order to combine quality characteristics for specific maximum performance. A broadly available, natural composite material resource is wood-defined as a natural polymeric, cellular fiber composite with superior advantages compared to other engineering materials [1]. Various wood-based products, such as cross laminated timber (CLT), particleboard, oriented strand boards (OSB) or fiberboards, resolve solid wood disadvantages, respectively anisotropy, biodegradability, and dimensional limitations [2].

Plywood, with its multilayered veneer-based laminar structure, is considered to be the oldest and the most important wood-based composite material with two distinct fields of application—construction purposes and multidimensional forming for interior and exterior products. Plywood can be bent by steaming a panel before forming, thin panels can be threaded and glued together or veneer layers are mechanically modified when producing 3D-molded parts [3]. To enhance veneer-based products (plywood and laminated veneer lumber) mechanical load bearing capacity, several experimental studies were conducted addressing synthetic glass, carbon, or other artificial fiber reinforcement [4–8]. Results revealed significantly improved mechanical properties for modulus of elasticity (MOE) and modulus of rupture (MOR), dimensional stability, and especially splitting strength [9–11]. Currently, the focus on fiber reinforcement shifted to more ecofriendly, fiber sources based on renewable resources. Flax (*Linum usitatissium*), ramie (*Boehmeria nivea*), hemp (*Cannabis sativa*), sisal (*Agave sisalana*), kenaf (*Hisbiscus cannabinus*), jute (*Corhorus capsularis*) or bamboo are characterized by biodegradability, cost effectiveness, natural availability [12,13]. Natural fibers are emerging at low cost, lightweight and apparently environmentally superior alternatives to glass fibers in composites. The current mass-based price of raw processed glass fibers with a density of 2.6 g/cm^3 is approximately 2 to 4 times greater than that of e.g., flax fibers with a density of 1.5 g/cm^3 [14,15]. The elevated mechanical properties of lightness, stiffness to weight ratio, and high tensile strength of natural fibers are advantageous for high performance composites [16,17]. For instance, flax with a density of 1.50 g/cm^3, tensile strength between 345 to 1100 N/mm^2, and an MOE of 27,600 to 64,500 N/mm^2 [18] is characterized by a relatively high stiffness, load capacity, and damping performance and shows comparable mechanical performance as e-glass and s-glass fibers. Moreover, flax is flexible and preferred for complicated geometries [19]. Böhm et al. [20] evaluated the bending characteristics of sandwich composite materials based on balsa plywood reinforced with flax (2 × 2 twill, 400 g/m^2) and glass fibers (roving fabric, 390 g/m^2) with epoxy resin. Reinforcement of the balsa plywood with both types of fibers significantly increased the modulus of elasticity and bending strength. Whilst for flax fibers, the flexural modulus was 7.5% higher than for glass fibers, the bending strength values of glass fibers were 26.5% lower compared to flax fibers. Research confirmed that flax fibers with less environmental impact may be a material that is used as an adequate replacement of glass fibers. Papadopoulos and Hague [21] prepared single-layer particleboards made from various wood chip/flax mixtures bonded with urea formaldehyde resin. The strength properties of boards containing up to 30% (mass-based) flax particles meet the minimum European standard (EN 310, 317, 319) requirements for interior particleboards. In the research of Susainathan et al. [22,23], glass fiber, carbon fiber, and flax fiber were used as the surface layer, poplar and okoume as the core layer to make wood-based sandwich structures. The impact and bending properties of the boards with flax fiber reinforcement showed comparable bending strength and stiffness as glass fiber reinforced samples. Research on fiber reinforcement primarily focuses on two-dimensional panel applications [24–26].

Bendability (formability) of two-dimensional curvature for wood is well investigated [27]. Three-dimensional (3D) molding of wood is considered as one of the most complicated chipless methods of shaping wood [28,29]. In general, limitations for veneer formability are set by the anisotropic character of wood and influenced by the MOE and tensile strength for the two directions parallel and perpendicular to grain. Wagenführ et al. [30] revealed a correlation between bending radius and veneer thickness for two-dimensional bending with a significant increase for the minimal possible radius with increasing veneer thickness and dependence on grain direction for non-modified veneers. Additional research on non-modified two-dimensional veneer bending suggested a higher suitability of tangential veneers for molding [31]. 3D veneer moldability is furthermore limited by the anisotropic nature, varying tensile strength, and plastic deformation of veneers [30]. Other relevant parameters are wood species, moisture content, sample and molding geometry in this respect. Deciduous wood species like beech (*Fagus sylvatica* L.) are considered more suitable for 3D-molding due to wood anatomy and plasticizing ability. Wood moisture content, heat pressing, moistening of wood are process variables which are relevant for surface quality in 3D wood molding [32]. It was shown that the 3D formability of veneer can be improved by 12 to 50% by increasing veneer moisture content and steaming. Birch (*Betula pendula Roth*) veneer showed better forming properties than beech (*Fagus sylvatica* L.) and ash (*Fraxinus excelsior*) [33]. Warping and cracking in the forming process depends strongly on veneer thickness [29]. According to Fekiač and Gáborík [27], there is no clear preference between radial and tangential veneers for 3D molding. Another way to increase the moldability of the wood is surface densification, hydrothermal plasticizing, thermal modification, reinforcement by natural fibers, and the optimization of the pressing process [34]. Zerbst et al. [35] successfully applied the Nakjima test to evaluate the deep drawing capacity of veneer laminate, which can be used for forming simulations. The authors found that cracks in veneer laminate sheets occur in the early wood zone and propagate in grain direction.

The aim of the study is to determine the influence of woven natural flax fiber reinforcement on 3D molded beech veneer-based plywood with a specific molding geometry. The occurrence of surface cracks and delamination as well as 3D-bending behavior for maximum load and MOE with the factors mold geometry and fiber-reinforcement are investigated.

2. Materials and Methods

2.1. Sample Preparation

Pre-conditioned (20 °C, 65% relative humidity) radial sliced zero defect beech (*Fagus sylvatica* L.) veneers with a thickness of 0.6 mm, an average density of 0.67 g/cm^3, and an average moisture content of 10.2 (SD = 0.3) % were used as wooden raw material in this study due to its superior bending strength and MOE performance compared to birch (*Betula pendula Roth*). Twill flax fabric LINEO FlaxPly Balanced Fabric 200 (Ecotechnilin, Valliquerville, France) with a thickness of 0.4 mm, a density of 1.27 g/cm^3, and a grammage of 200 g/m^2 acted as fiber reinforcement. The flax moisture content accounted for 11 (SD = 0.3) %. West Systems International (Romsey, England) 105 Epoxy Resin and 207 Special Coating Hardener were used as adhesive. Two kinds of lay-ups were introduced (Figure 1). The unreinforced reference samples existed of ninety degree cross layered veneer layers, whereas the flax fiber reinforced samples consisted of the identical ninety cross layered veneer layers with additional four layers of flax fabric. These were located at the first and second glue line on each side in order to improve the tensile strength under bending and to minimize the influence of shear stresses. The calculated amount of epoxy resin per glue line was set to 200 g/m^2 for the veneer to veneer layers and 400 g/m^2 for flax fiber to veneer.

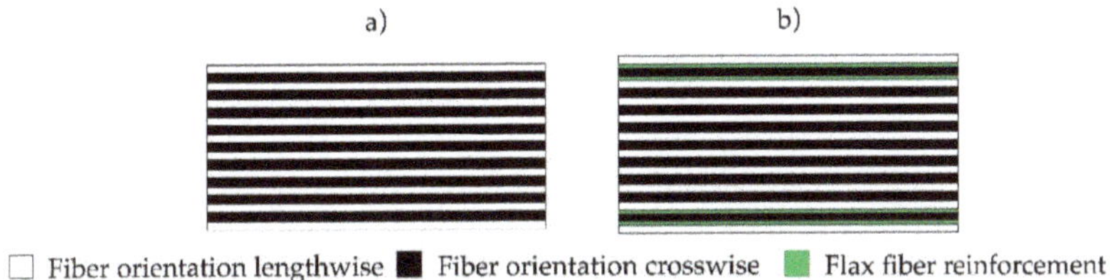

Figure 1. Veneer layup reference (**a**), with flax fiber reinforcement (**b**).

The specimens were designed focusing on the use of the molded parts as car seat shells (Figure 2). A transition section between seating and back was considered in the design of the specimens. The seating shell was split in the center, and looking in driving direction, the left part of the shell was considered. The size of the raw veneer sheets accounted for 20 × 27.5 cm.

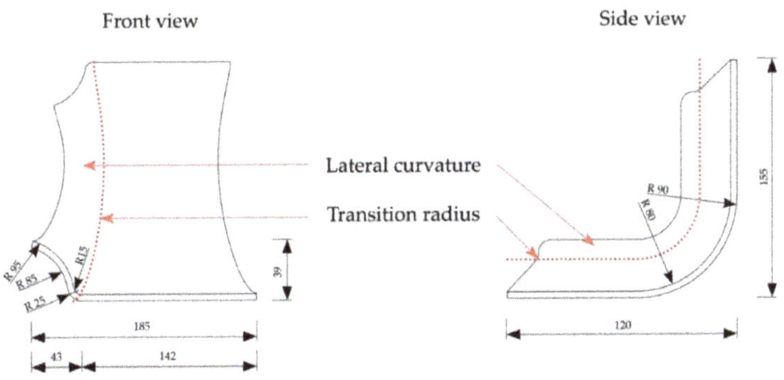

Figure 2. Dimension of the 3D molded specimens (dimensions in mm).

To overcome the veneer warping during molding, two different options where used. First, a minimal leaving out of the critical curvature was applied (Figure 3).

a) b)

Figure 3. Cut-out of specimens (**a**) reference, (**b**) with flax fiber reinforcement.

Second, relief cuts at the critical curvature were provided (Figure 4). These were filled with an Epoxy filler (Molto 2-components wood replacement) after pressing. In total, four different types of setups with three samples each were tested.

a) b)

Figure 4. Relief cut of specimens (**a**) reference, (**b**) with flax fiber reinforcement.

The lay-up of the veneer laminates and the glue application were conducted manually. The calculated areal glue amount spreading per layer was checked by a scales KERN PRS 620-6 (Kern & Sohn GmbH, Ballingen-Frommern, Germany). The controlled molding was carried out using a HÖFLER (Taiskirchen, Austria) HLOP 280 press to reach the target thickness of 10 mm within 15 h of cold pressing. The applied pressure during molding was 2.5 N/mm^2 for the reference samples and 2.8 N/mm^2 for the flax fiber reinforced samples. The temperature was kept constant at 20 °C. No pretreatment of the veneers was applied to improve the bending performance. Before further testing, the samples were stored for seven days under constant standard climate conditions (20 °C, 65% relative humidity).

2.2. Testing of Surface Cracks and Delamination

Surface cracks and delamination were determined after pressing and conditioning of the formed parts. Samples were trimmed to equal size before proceeding to the 3D bending test.

To determine the surface crack length and widths, pictures were taken with a digital camera (Rollei Compactline 750) and a scale positioned on the formed parts. For the exact determination of length and width, MATLAB R2018a was used according to Equation (1). On each picture taken, a scale was installed, to determine the dimensions of cracks and delamination considering the pixel-based length of the defect l_{pixel}, the pixel-based length of the reference scale $l_{ref.\,pixel}$, and the known length of the scale $l_{ref.}$ The delamination length was determined using the identical formula. In addition, a verbal nomenklatura was introduced to describe the location of surface cracks and delamination in order to locate critical spots of failure. First of all, cracks and delamination were categorized whether they occurred (1–1) on the front or (1–2) back of the formed part. Second, it was described whether they occurred (2–1) on the back shell or (2–2) on the seat. Third, defects were assessed regarding their position relative to the transition radius; (3–1) transition radius, (3–2) above transition radius, (3–3) below transition radius. Finally, defects on edges were assessed using the attributes (4–1) outer edge, (4–2) edge cut-out. All measurements were taken from formed parts to exclude influences caused by post-processing.

$$l = \frac{l_{pixel}}{l_{ref.\,pixel}} \times l_{ref.} \tag{1}$$

2.3. Three-Dimensional Bending Test

A modified bending test following UN/ECE Regulations Nr. 17 [36] with a variation regarding the direction of force for spreading was implemented. For testing, the samples were placed with the small faces on an aluminum block due to its low friction coefficient to reduce influences on internal stress distribution (Figure 5). Constant linear force with a speed of 10 mm/min was applied by a pressure disc (ø 135 mm) parallel to the supports using a Zwick/Röll Z 250. The maximum load (F_{max}) and the modulus of elasticity (MOE) were determined following EN 310:2005 [37].

Figure 5. Bending test for 3D molded plywood samples.

The statistical significance of the factors molding geometry and fiber-reinforcement was determined using two-way ANOVAs with consideration of interaction effects of first order.

3. Results and Discussion

3.1. Number of Surface Cracks and Delamination

The finished specimens showed up to 6 cracks and between 1 and 2 delamination after pressing. Comparing the number of counted surface cracks for each of the four groups, it was assumed that there is an influence of the factors molding geometry and fiber reinforcement. Based on a two-way ANOVA to prove the statistical significance of the factors molding geometry (*p*-value 0.074) and fiber reinforcement (*p*-value 0.334) displayed that there is no significant influence. In addition, there is no significant interaction effect of the factors. This can be explained by the high variation of crack numbers between specimens (Table 1).

Table 1. Number of cracks and crack geometry after bending (SD in brackets).

Specimen	No. of Cracks	Crack Length	Crack Width	N
		(mm)	(mm)	
Cut-out	1 (2)	18.2 (7.0)	0.6 (0.4)	3
Cut-out reinf.	1 (1)	6.7 (2.3)	0.4 (0.2)	3
Relief cut	2 (1)	12.2 (3.9)	0.4 (0.1)	3
Relief cut reinf.	4 (2)	17.4 (11.1)	0.5 (0.3)	3

To determine the influence of the factors molding geometry and fiber reinforcement on the number of delamination, the same procedure as for the surface cracks was applied. The two-way ANOVA displayed a significant influence of the molding geometry with a p-value of 0.037. On average, formed parts with a cut-out showed one delamination, whilst parts with relief cuts had two delaminations after pressing (Table 2). The fiber-reinforcement (p-value 0.631) such as the factor interaction (p-value 0.631) in contrast had no significant influence on the number of delaminations.

Table 2. Number and length of delamination after bending (SD in brackets).

Specimen	No. of Delamination	Delamination Length	N
		(mm)	
Cut-out	1 (1)	12.7 (4.8)	3
Cut-out reinf.	1 (1)	8.9 (3.4)	3
Relief cut	2 (0)	12.3 (3.3)	3
Relief cut reinf.	2 (0)	11.3 (3.4)	3

Considering these statistics, it seems that the cracks and delamination are predominantly caused by deformations occurring during the molding process rather than ineffective gluing between composite layers.

3.2. Length and Width of Surface Cracks and Delamination

The effect of the factors molding geometry and fiber reinforcement on the length of cracks was assessed using a two-way ANOVA. Only specimens with cracks were considered. According to the ANOVA, the p-value for the interaction between the factors is significant with 0.030, stating the fact that the level effect of one factor is dependent on the level of the other factor. Cracks on fiber reinforced parts with cut-out have an average length of 6.7 (SD = 2.3) mm, 64% lower than their unreinforced counterpart—a coherence that could not be shown for formed parts with relief cuts.

In contrast, the results of the surface crack widths had no statistically significant factor effect or interaction (p-value 0.196) between the two factors molding geometry and fiber reinforcement. This is underlined by the singular p-value for molding geometry with 0.774 and fiber reinforcement with a p-value 0.864.

Based on the comparison of the mean and standard deviations of the factors, the two-way ANOVA revealed no statistically significant influence on the delamination length. Molding geometry (p-value 0.587) and fiber reinforcement (p-value 0.203) showed no significant interaction effect (p-value of 0.450).

The sum of the crack length of a specimen is significantly (p-value 0.034) affected by the molding geometry, showing that it accounts for 16.6 (SD = 27.9) mm with formed parts with cut-out compared to 64.6 (SD = 33.5) mm with relief cut. The cumulated crack width per sample is not significantly affected by the mold geometry or fiber reinforcement. The effect of molding geometry on the cumulated length of delamination is narrowly not significant (p-value 0.053). The average sum of delamination length accounts for 12.9 (SD = 10.5) mm when using cut-outs and 23.6 (SD = 4.0) mm when relief cuts are applied (Table 3). Summarized, it was found that molded parts with relief cut have more and additionally more significant cracks and delamination than parts with a cut-out. The sum of the crack lengths is two thirds lower when applying fiber reinforcement for parts with cut-out, whereas no advantage of fiber reinforcement could be detected for parts with relief cuts. The investigation of cracks and delamination shows that the total sum of defect characteristics is higher for parts with relief cuts. This is due to the fact that a greater part of the critical sector of the transition zone is covered. The forced deformations are greater with the relief cut formed parts and exceed more often the strength of the veneer. The fiber reinforced parts with cut-out show lower crack and delamination length, probably due to the fact that fabric with twill weave is easily shear formable and has good draping properties [38].

Table 3. Number and length of delamination after bending (SD in brackets).

Specimen	Sum Crack Length	Sum Crack Width	Sum Delam. Length	N
	(mm)	(mm)	(mm)	
Cut-out	24.2 (41.9)	0.6 (1.0)	16.9 (13.0)	3
Cut-out reinf.	8.9 (4.1)	1.1 (0.9)	8.9 (7.8)	3
Relief cut	53.9 (38.5)	0.6 (0.4)	24.6 (5.4)	3
Relief cut reinf.	75.3 (31.3)	2.2 (1.4)	22.6 (2.8)	3

3.3. Location and Orientation of Defects

Surface cracks occurred with a relative frequency of 67 (SD = 31) % on the inner (positive) side of the molding with the shorter radius. Ninety-seven (SD = 11) percent are located at the "transition radius" due to the sharp radius and the change of direction into three different dimensions. Seventy-six (SD = 22) percent are placed on the outer edges of the molding (Figure 6). The reason is, that tensions cannot be transferred to neighboring veneer layers in the corner range which results in cracks. All surface cracks are oriented in grain direction, because the tensile strength of beech veneer orthogonally to grain direction is 20 times lower than in grain direction [39], which is confirmed by Zerbst et al. [35] finding the same phenomenon when producing molded veneer laminates.

Cracks		Cut-out			Cut-out reinf.			Relief cut			Relief cut reinf.		
Sample		1	2	3	4	5	6	7	8	9	10	11	12
Total number of cracks		4	0	0	2	1	1	3	1	3	2	6	5
Global position	(1-1) Front	2	0	0	2	1	1	1	1	2	1	3	1
	(1-2) Back	2	0	0	0	0	0	2	0	1	1	3	4
Detailed position	(2-1) Back shell	1	0	0	0	1	0	3	1	3	1	5	4
	(2-2) Seat shell	3	0	0	2	0	1	0	0	0	0	0	0
Position on transition radius	(3-1) Transition radius	4	0	0	2	1	1	3	1	2	2	6	5
	(3-2) Above transition radius	0	0	0	0	0	0	0	0	1	0	0	0
	(3-3) Below transition radius	0	0	0	0	0	0	0	0	0	0	0	0
Position on edges	(4-1) Outer edge	2	0	0	2	1	1	2	1	2	1	4	3
	(4-2) Edge with cut-out	2	0	0	0	0	0						

Figure 6. Heat map of crack number and position.

Cracks are not acceptable in molded parts due to structural and aesthetic reasons. A possible optimization is to create micro cracks and collapsed cells in veneer layers before molding to prevent the formation of macro cracks. It is important to note that a collapsed cell not necessarily goes along with a broken cell wall [30].

Delaminations were predominantly found on the inner (front) side as well (Figure 7). Seventy-seven (SD = 26) percent occurred on the more strongly bent inner side. All of them were found above the transition radius on the formed parts where the shear strengths were strongest. For the same reason, these occurred in large part with some distance to the cut-out referenced to the edge position. Ninety-five percent of all delaminations were found near the outer edge of the formed parts. The use of rigid press forms results in an uneven pressure distribution because the application of uniaxial vertical pressure results in uneven orthogonal pressure referred to the curved surface of the formed part [40]. The maximal angle of the form parts used in this study is 50 degrees on the outer edge and the pressure was therefore 36% [40] reduced in this area, as an explanation for the less efficient bonding. The strong influence of the mold geometry on the quality of plywood formed parts was discussed by Comsa [41], suggesting deformation modeling using finite element methods. This shows an interesting opportunity for the current mold part to lower crack and delamination formation during molding. Finally, the use of plasticized veneer (at least the outer layers) before pressing suggests to lower the risk for cracks [34].

Delaminations		Cut-out			Cut-out reinf.			Relief cut			Relief cut reinf.		
Sample		1	2	3	4	5	6	7	8	9	10	11	12
Total number of delaminations		2	1	1	2	1	1	2	2	2	2	2	2
Global position	(1-1) Front	2	1	1	0	1	2	1	2	1	1	1	1
	(1-2) Back	0	0	0	0	1	1	1	0	1	1	1	1
Detailed position	(2-1) Back shell	1	1	1	0	0	1	2	1	2	2	2	2
	(2-2) Seat shell	1	0	0	0	0	0	0	1	0	0	0	0
Position on transition radius	(3-1) Transition radius	0	0	0	0	0	0	0	0	0	0	0	0
	(3-2) Above transition radius	2	1	1	0	1	2	2	2	2	2	2	2
	(3-3) Below transition radius	0	0	0	0	0	0	0	0	0	0	0	0
Position on edges	(4-1) Outer edge	1	1	1	0	1	2	2	2	2	2	2	2
	(4-2) Edge with cut-out	1	0	0	0	0	0						

Figure 7. Heat map of delamination number and position in the 3D-molded plywood samples after forming.

3.4. 3D Bending Test

Based on the load-deformation curvature, two thirds of the samples with cut-outs showed a linear-elastic deformation behavior and a nondescriptive fracture with a mixture of delaminations and failures within the beech veneer (Figure 8). One third of samples had a partial fracture behavior within the linear-elastic deformation with a notable tension failure. The load-deformation curve for relief cut samples revealed a semi-fracture at the end of the linear-elastic range caused by the failure of the filler regardless of fiber reinforcement, followed by a further plastic deformation. Fiber reinforcement resulted in an increase in force for both cut-out and relief cut samples in the plastic deformation range, whereas unreinforced samples of both groups more or less stagnated. These findings validate research by Wagenführ et al. [30] considering that veneer laminates behave predominantly elastically.

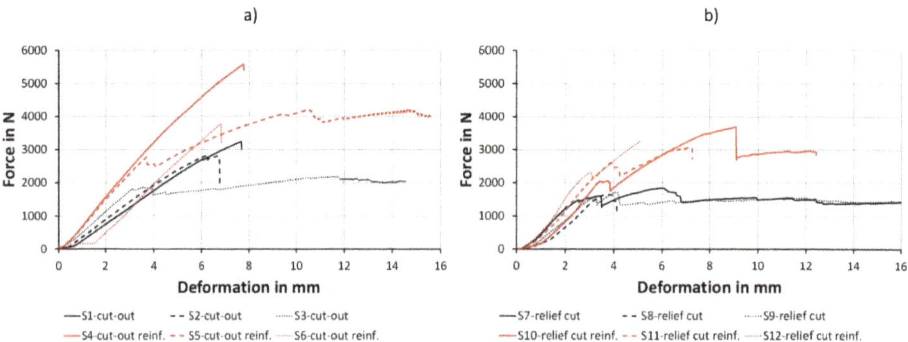

Figure 8. Force-path diagram for specimens with cut-out (**a**) and relief cut (**b**).

The results of maximum load displayed a significant influence of the molding geometry (*p*-value 0.010) as well as the fiber reinforcement (*p*-value 0.001). Moldings with cut-outs resulted in 58 and 36% higher bending strength than their counterparts with relief cuts. Fiber reinforcement lead to 65 and 92% higher bending strength compared with the unreinforced part, respectively (Table 4).

Table 4. Results of the bending test (SD in brackets).

Specimen	Max. Force	MOE	N
	(N)	(N/mm^2)	
Cut-out	2784 (530)	532 (45)	3
Cut-out reinf.	4531 (941)	791 (68)	3
Relief cut	1736 (97)	633 (70)	3
Relief cut reinf.	3340 (322)	816 (97)	3

The reason for the significant increase in the maximal load capacity through flax fiber reinforcement is the high tensile strength of flax fibers, approximately 5 times as high as the one of the wood fiber [38]. This increased tensile strength affects especially the inner side of the formed part which is tensile-loaded in the current experimental situation. As the relief cuts are an interruption in the fiber reinforcement, the filler grouts fail early and are an explanation for the lower performance of such parts.

Regarding the results of the MOE, similar effects are given. The influence of the fiber reinforcement is highly significant (p-value 0.001), whereas the influence of the molding geometry—in contrast to the maximum load capacity—was statistically not significant (p-value 0.168). Formed parts which are fiber reinforced had an MOE on average 76% higher than the unreinforced parts (Table 4). The mold geometry does not have an effect in this respect, because failures occur above 40% of the maximal force and up to this point the part shows linear-elastic behavior.

A significant correlation between the sum of crack width in a part and the MOE was observed (p-value 0.024, R^2 = 0.41). The higher the cumulated crack width, the higher the MOE. Moreover, a higher number of cracks result in a not significant higher MOE. A rationale for this coherence is that crack formation during the forming process results in an improved layer contact due to a compensation of the deformations in the transition zones.

4. Conclusions

The three-dimensional bending tests regarding the maximum load capacity and the MOE showed that the performance of beech veneer-based 3D molded plywood could be significantly improved by flax fiber reinforcement. In addition, the influence of the mold geometry is significant for the load capacity, but not for the MOE.

Formed parts with few and less pronounced defects, such as cracks and delamination, can be produced using cut-outs in the most deformed area of the 3D molded parts. The location of cracks and delamination found in this study strongly suggests the optimization of the press form and the application of multiaxial press processes to avoid delamination. Moreover, 3D molded parts should be produced with some oversize as 68% of the cracks and 95% of the delamination was found on the outer edges. A majority of defects could be removed in a final trimming process.

Further research is recommended for a deeper understanding of the interaction and influence of veneer thickness and flax fabrics grammage on formability, as well as the influence of veneer pretreatment and hot pressing to improve 3D formability. Additional research should focus on interactions between veneer, flax, and adhesive to determine a better understanding of delamination resistance and supplemented by an evaluation of fiber wetting resulting from the resin. Cost and production aspects have to be taken into account for broader industrial applications. Especially the research on cost efficient industrial glue systems, glue amount optimization, and the influence on the performance of woven flax fiber fabric is worth further research.

Author Contributions: Conceptualization and methodology, J.J., G.K. and M.H.; validation and formal analysis, J.J., G.K. and M.-C.B.; investigation, J.J., G.K. and M.H.; resources and funding acquisition, M.-C.B. and L.K.; writing—original draft preparation, review and editing, J.J., G.K., M.-C.B. and L.K.; supervision and project administration, G.K., M.-C.B. and L.K. All authors have read and agreed to the published version of the manuscript.

Funding: This research was supported by Austrian Founding Agency (FFG) Innovations Plus No. 866425 and by the Slovak Research and Development Agency under contracts no. APVV-18-0378, APVV-19-0269, and VEGA 1/0717/19.

Acknowledgments: The authors are thankful for the support of J.B. and F.P. for their support in fabricating the press forms and gratefully acknowledge the support of Ing. Thomas Wimmer for his contribution in conducting mechanical testing at Salzburg University of Applied Sciences at Campus Kuchl.

Conflicts of Interest: The authors declare no conflict of interest.

References

1. Mahut, J.; Reh, R. *Plywood and Decorative Veneers*; Technical University of Zvolen: Zvolen, Slovakia, 2007.
2. Stark, N.M.; Cai, Z.; Carll, C. Chapter 11—Wood-Based-Composite Materials and Panel Products, Glued Laminated Timber, Structural Materials. In *Wood Handbook—Wood as an Engineering Material*; U.S. Department of Agriculture, Forest Service, Forest Products Laboratory: Madison, WI, USA, 2010.
3. Panic, L.; Hodzic, A.; Nezirevic, E. Modern and sophisticated processes of 3D veneer plywood bending. *Acta Tech. Corviniensis Bull. Eng.* **2016**, *9*, 2067–3809.
4. Muthuraj, R.; Misra, M.; Defersha, F.M.; Mohanty, A.K. Influence of processing parameters on the impact strength of biocomposites: A statistical approach. *Compos. Part A Appl. Sci. Manuf.* **2016**, *83*, 120–129. [CrossRef]
5. Percin, O.; Altunok, M. Some physical and mechanical properties of laminated veneer lumber reinforced with carbon fiber using heat-treated beech veneer. *Holz Roh Werkst.* **2017**, *75*, 193–201. [CrossRef]
6. Liu, H.; Luo, B.; Shen, S.; Liu, H. Design and mechanical tests of basalt fiber cloth with MAH grafted reinforced bamboo and poplar veneer composite. *Holz Roh Werkst.* **2018**, *77*, 271–278. [CrossRef]
7. Auriga, R.; Gumowska, A.; Szymanowski, K.; Wronka, A.; Robles, E.; Ocipka, P.; Kowaluk, G. Performance properties of plywood composites reinforced with carbon fibers. *Compos. Struct.* **2020**, *248*, 112533. [CrossRef]
8. Liu, Y.; Guan, M.; Chen, X.; Zhang, Y.; Zhou, M. Flexural properties evaluation of carbon-fiber fabric reinforced poplar/eucalyptus composite plywood formwork. *Compos. Struct.* **2019**, *224*, 111073. [CrossRef]
9. Xu, H.; Nakao, T.; Tanaka, C.; Yoshinobu, M.; Katayama, H. Effects of fiber length and orientation on elasticity of fiber-reinforced plywood. *J. Wood Sci.* **1998**, *44*, 343–347. [CrossRef]
10. Rowlands, R.E.; Deweghe, R.P.; Laufenberg, T.L.; Krueger, G.P. Fiber-reinforced wood composites. *Wood Fiber Sci.* **1986**, *18*, 39–57.
11. Bal, B.C.; Bektaş, I.; Mengeloğlu, F.; Karakuş, K.; Demir, H.Ö. Some technological properties of poplar plywood panels reinforced with glass fiber fabric. *Constr. Build. Mater.* **2015**, *101*, 952–957. [CrossRef]
12. Sorieul, M.; Dickson, A.R.; Hill, S.J.; Pearson, H. Plant Fibre: Molecular Structure and Biomechanical Properties, of a Complex Living Material, Influencing Its Deconstruction towards a Biobased Composite. *Materials* **2016**, *9*, 618. [CrossRef]
13. Ticoalu, A.; Aravinthan, T.; Cardona, F. A Review of Current Development in Natural Fiber A Review of Current Development in Natural Fiber Composites for Structural and Infrastructure Applications. In Proceedings of the Southern Region Engineering Conference, Toowoomba, Australia, 11–12 November 2010.
14. Šedivka, P.; Bomba, J.; Böhm, M.; Zeidler, A. Determination of Strength Characteristics of Construction Timber Strengthened with Carbon and Glass Fibre Composite Using a Destructive Method. *Bioresources* **2015**, *10*, 4674–4685. [CrossRef]
15. Joshi, S.; Drzal, L.; Mohanty, A.; Arora, S. Are natural fiber composites environmentally superior to glass fiber reinforced composites? *Compos. Part A Appl. Sci. Manuf.* **2004**, *35*, 371–376. [CrossRef]
16. Borri, A.; Corradi, M.; Speranzini, E. Reinforcement of wood with natural fibers. *Compos. Part B Eng.* **2013**, *53*, 1–8. [CrossRef]
17. Sam-Brew, S.; Smith, G. Flax and Hemp fiber-reinforced particleboard. *Ind. Crops Prod.* **2015**, *77*, 940–948. [CrossRef]
18. Mohanty, A.K.; Misra, M.; Hinrichsen, G. Biofibres, biodegradable polymers and biocomposites: An overview. *Macromol. Mater. Eng.* **2000**, *276*, 1–24. [CrossRef]
19. Goudenhooft, C.; Bourmaud, A.; Baley, C. Flax (*Linum usitatissimum* L.) Fibers for Composite Reinforcement: Exploring the Link between Plant Growth, Cell Walls Development, and Fiber Properties. *Front. Plant Sci.* **2019**, *10*, 411. [CrossRef]

20. Böhm, M.; Brejcha, V.; Jerman, M.; Černý, R. Bending Characteristics of Fiber-Reinforced Composite with Plywood Balsa Core. In Proceedings of the International Conference of Computational Methods in Sciences and Engineering 2019 (ICCMSE-2019), Rhodes, Greece, 1–5 May 2019; Volume 2186, p. 070006.

21. Papadopoulos, A.N.; Hague, J.R. The potential for using flax (*Linum usitatissimum* L.) shiv as a lignocellulosic raw material for particleboard. *Ind. Crops Prod.* **2003**, *17*, 143–147. [CrossRef]

22. Susainathan, J.; Eyma, F.; De Luycker, E.; Cantarel, A.; Castanié, B. Experimental investigation of impact behavior of wood-based sandwich structures. *Compos. Part A Appl. Sci. Manuf.* **2018**, *109*, 10–19. [CrossRef]

23. Susainathan, J.; Eyma, F.; De Luycker, E.; Cantarel, A.; Castanier, B. Manufacturing and quasi-static bending behavior of wood-based sandwich structures. *Compos. Struct.* **2017**, *182*, 487–504. [CrossRef]

24. Mathijsen, D. The renaissance of flax fibers. *Reinf. Plast.* **2018**, *62*, 138–147. [CrossRef]

25. Prabhakaran, S.; Krishnaraj, V.; Sharma, S.; Senthilkumar, M.; Jegathishkumar, R.; Zitoune, R. Experimental study on thermal and morphological analyses of green composite sandwich made of flax and agglomerated cork. *J. Therm. Anal. Calorim.* **2019**, *139*, 3003–3012. [CrossRef]

26. Jorda, J.S.; Barbu, M.C.; Kral, P. Natural fiber reinforced veneer based products. *Pro Ligno* **2019**, *15*, 206–219.

27. Fekiac, J.; Gáborík, J. *Formability of Radial and Tangential Beech Veneers*; Annals of Warsaw University of Life Sciences: Warsaw, Poland, 2016; pp. 191–197.

28. Wagenführ, A.; Buchelt, B. Untersuchungen zum Materialverhalten beim dreidimensionalen Formen von Furnier. *Holztechnologie* **2005**, *46*, 13–19.

29. Gaff, M.; Gašparík, M. 3D Molding of Veneers by Mechanical and Pneumatic Methods. *Materials* **2017**, *10*, 321. [CrossRef]

30. Wagenführ, A.; Buchelt, B.; Pfriem, A. Material behaviour of veneer during multidimensional moulding. *Holz Roh Werkst.* **2005**, *64*, 83–89. [CrossRef]

31. Langova, N.; Joscak, P.; Mozuchova, M.; Trencanova, L. Analysis the effects of bending load of veneers for purposes of planar moulding. *Ann. Wars. Univ. Life Sci.* **2013**, *83*, 173–178.

32. Gaff, M.; Gáborík, J. Evaluation of Wood Surface Quality after 3D Molding of Wood by Pressing. *Bioresources* **2014**, *9*, 4468–4476. [CrossRef]

33. Fekiac, J.; Gáborík, J.; Smidriakova, M. 3D formability of moistened and steamed veneers. *Acta Fac. Xylologiae Zvolen* **2016**, *58*, 15–26.

34. Zemiar, J.; Fekiac, J.; Gaborik, J.; Petro, A. Three-dimensional formability of rolled, pressed, and plasticized veneers. *Ann. Wars. Univ. Life Sci.* **2013**, *84*, 339–343.

35. Zerbst, D.; Affronti, E.; Gereke, T.; Buchelt, B.; Clauß, S.; Merklein, M.; Cherif, C. Experimental analysis of the forming behavior of ash wood veneer with nonwoven backings. *Holz Roh Werkst.* **2020**, *78*, 321–331. [CrossRef]

36. United Nations Economic Commission for Europe. Regulation No 17 of the Economic Commission for Europe of the United Nations (UN/ECE)—Uniform Provisions Concerning the Approval of Vehicles with Regard to the Seats, Their Anchorages and Any Head Restraints. Available online: https://op.europa.eu/en/publication-detail/-/publication/4d5ab93c-7d45-4b3a-b49f-b10b6476b5df (accessed on 29 October 2020).

37. European Committee for Standardization. *EN 310:2005 Wood Based Panels—Determination of Modulus of Elasticity in Bending and of Bending Strength*; European Committee for Standardization: Brussels, Belgium, 2005.

38. Schürmann, H. *Konstruieren Mit Faser-Kunststoff-Verbunden*, 2nd ed.; Springer: Berlin, Germany, 2007.

39. Wagenführ, R. *Holzatlas*; Carl Hanser Verlag GmbH & Co. KG: München, Germany, 2006.

40. Kollmann, F. *Technologie des Holzes und der Holzwerkstoffe*; Springer: Berlin/Heidelberg, Germany, 1955.

41. Comsa, G.N. Dimensional and geometrical optimization of structures and materials for curved or molded chair furniture. In Proceedings of the 3rd International Conference on Advanced Composite Materials Engineering COMAT, Brasov, Romania, 27–29 October 2010.

Publisher's Note: MDPI stays neutral with regard to jurisdictional claims in published maps and institutional affiliations.

polymers

Article

The Evaluation of Torrefied Wood Using a Cone Calorimeter

Peter Rantuch [1,*] , Jozef Martinka [1] and Aleš Ház [2]

1   Faculty of Materials Science and Technology in Trnava, Slovak University of Technology in Bratislava, 917 24 Trnava, Slovakia; jozef.martinka@stuba.sk
2   Faculty of Chemical and Food Technology, Slovak University of Technology in Bratislava, 812 37 Bratislava, Slovakia; ales.haz@stuba.sk
*   Correspondence: peter.rantuch@stuba.sk; Tel.: +421-910-993-650

Abstract: This study focuses on the energy potential and combustion process of torrefied wood. Samples were prepared through the torrefaction of five types of wood: Ash, beech, oak, pine and spruce. These were heated for 2 h at a temperature of 300 °C under a nitrogen atmosphere. Torrefied wood was prepared from wood samples with dimensions of $100 \times 100 \times 20$ mm^3. These dimensions have enabled investigation of torrefied wood combustion in compact form. The effect of the external heat flux on the combustion of the samples was measured using a cone calorimeter. The observed parameters, include initiation times, heat release rate and combustion efficiency. The results show that increasing the external heat flux decreases the evenness of combustion of torrefied wood. At the same time, it increases the combustion efficiency, which reached an average value of approximately 72% at 20 kW m^{-2}, 81% at 30 kW m^{-2} and 90% at 40 kW m^{-2}. The calculated values of critical heat flux of the individual samples ranged from 4.67 kW m^{-2} to 15.2 kW m^{-2}, the thermal response parameter ranged from 134 kW s$^{0.5}$ m^{-2} to 297 kW s$^{0.5}$ m^{-2} and calculated ignition temperature ranged from 277 °C to 452 °C. Obtained results are useful both for energy production field and for fire safety risk assessment of stored torrefied wood.

Keywords: torrefied wood; fuel; combustion; heat release rate

Citation: Rantuch, P.; Martinka, J.; Ház, A. The Evaluation of Torrefied Wood Using a Cone Calorimeter. *Polymers* **2021**, *13*, 1748. https://doi.org/10.3390/polym13111748

Academic Editor: Antonios Papadopoulos

Received: 11 May 2021
Accepted: 23 May 2021
Published: 27 May 2021

Publisher's Note: MDPI stays neutral with regard to jurisdictional claims in published maps and institutional affiliations.

1. Introduction

The current way in which natural fossil resources are consumed to provide energy does not reflect the concept of sustainability [1]. Sustainable development is development that meets the needs of the present without compromising the ability of future generations to meet their own needs [2]. Therefore, the importance of renewable energy sources is growing. One of the possible solutions may be a more efficient use of biomass. It is a primary source of renewable carbon that can be utilised as a feedstock for biofuels or biochemical production in order to achieve energy independence [3].

In 2015, the worldwide total primary energy supply was 13,647 Mtoe, of which 13.4%, or 1823 Mtoe, came from renewable energy sources. Due to its widespread non-commercial use in developing countries, solid biofuels/charcoal remains the largest renewable energy source, representing 63.7% of the global renewable supply [4]. Torrefied wood is a fuel with the potential to partially replace coal [5].

Torrefaction is a pyrolysis process carried out at a temperature range of 200 to 300 °C under an inert atmosphere, which produces a high-quality solid biofuel that can be used for combustion and gasification [3,6,7]. It removes moisture and low weight organic volatile components and depolymerises the long polysaccharide chains, producing a hydrophobic solid product with an increased energy density (on a mass basis) and greatly increased grindability [8].

Hemicellulose, cellulose, and lignin are the basic constituents of a biomass and their thermal behaviour is highly related to the degradation of the biomass in a high-temperature environment. Biomass with torrefaction temperatures of 200 to 225 °C are described as light torrefaction; 250 °C as mild torrefaction, and 275 to 300 °C belong to severe torrefaction [6].

Based on the thermal analysis results Chen and Kuo stated that xylan is always sensitive to torrefaction in the temperature range of 200 to 300 °C. As the torrefaction temperature is no higher than 225 °C, weight loss of hemicellulose is very low. This temperature, thus, plays no part in thermal degradation of hemicellulose. Thermal degradation of cellulose is slight if the torrefaction temperature is less than or equal to 250 °C [6]. By decomposing the reactive hemicellulose fraction, a fuel with increased energy density is produced. [9] Simultaneously part of oxygen is removed from biomass [7]. During the process of wood torrefaction and with increasing temperature and time of exposure, the amount of fixed carbon, lignin and carbon in the end product also rises, with temperature as the most important factor [10]. The liquid yield is also increased [11]. According to Wannapeera and Worasuwannarak torrefaction conditions have impact on the elemental composition of torrefied wood only at a higher mass yield (>80%). Energy yield decreases with increasing degree of torrefaction [12].

The main gaseous products of the torrefied biomass combustion process are CO_2 and H_2O which confirms that carbon and hydrogen are significant compounds in torrefied biomass. The amount of gas decreases with increasing torrefaction temperature, probably because of gas removal during the torrefaction process [13]. Torrefied biomass has a higher pyrolysis and combustion temperature due to moisture and volatiles removal and thermal decomposition of its main components. Torrefaction also increases ash content and C/H and C/O ratio of biomass [14]. The increase in ash content of torrefied biomass is mainly due to mass loss during torrefaction reaction [15]. The lower O/C and H/C ratio is due to removal of water and carbon dioxide [16]. Moisure absorption of torrefied wood is significantly reduced due to loss of hydroxyl groups [15]. In 2019 there were produced 431 4342 t wastes from wood processing and the production of paper, cardboard, pulp, panels and furniture in the Slovak republic [17]. Although, this waste can be used for the production of other materials, such as eco-friendly, high-density fiberboards [18], 42.7% of this amount, was incinerated with energy recovery [16]. The most harvested wood in the Slovak Republic is spruce, followed by beech, fir and oak [19]. Based on data from 2019, beech (34.2%), spruce (22.1%), oak (10.5%) and pine (6.6%) have the highest proportion in forests in the Slovak Republic [20]. In relation to logging and tree species proportion, beech, oak, spruce and pine were selected as samples. Ash was chosen as a representative of less common species.

Although the torrefaction process has been used for a long time, it is still one of the important energy recovery options for biomass waste. In contrast to most previous works, torrefied wood was produced from bigger samples, not from disintegrated wood (this research represents the border between laboratory and medium-sized experiments). The burning of dust particles is significantly different from the burning of compact material. For example, in the case of dust cloud, explosive combustion can occur [21]. Our approach allowed the investigation of torrefied wood combustion in compact form, in the contrast to previously published works. The literature also lacks a description of torrefied wood in terms of its fire safety during storage. In these cases, it may be an additional fuel and may result in an increase of heat release rate during a fire. This factor subsequently affects the load-bearing capacity and integrity of the surrounding structures. Measurements using a conical calorimeter are suitable for such assessment of materials [22,23]. The aim of this article is to assess torrefied wood prepared from different woods using a cone calorimeter. The obtained results can be used in terms of energy recovery or for the needs of fire protection during storage.

2. Materials and Methods

Samples of five types of wood were selected for the preparation of torrefied wood. These included the wood of three deciduous trees: ash (*Fraxinus excelsior*), beech (*Fagus sylvatica*) and oak (*Quercus petraea*); and two coniferous trees: Spruce (*Picea abies*) and pine (*Pinus radiata*). The samples were cut tangentially into pieces with dimensions of 100 mm × 100 mm and 20 mm width. The schematic of the sample preparation device is shown in Figure 1. The torrefaction process was based on a method indicated by Liu et al. for the torrefaction of bamboo [24]. Nitrogen was used as the protective gas. It was continuously supplied to the muffle furnaceat a flow rate of 500 mL min^{-1}. The samples of wood were dried at 105 °C for 24 h, and then they were inserted into the heated Nabertherm Muffle Furnace L24/11/P330 (Nabetherm GmbH, Bremen, Germany) with the temperature set at 300 °C. The residence time was 2 h. After torrefaction, the samples were placed into a desiccator, where they cooled to the ambient temperature.

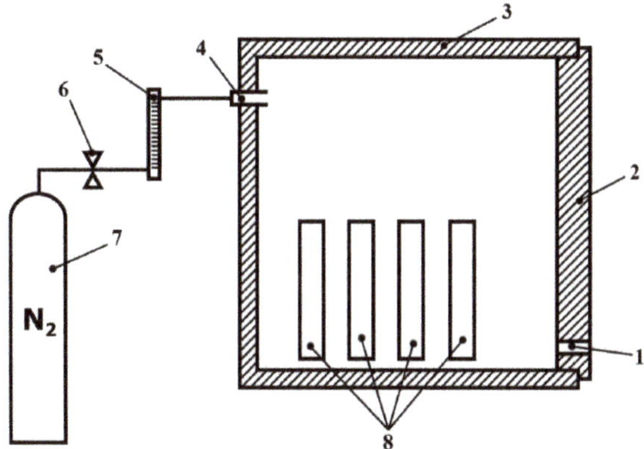

Figure 1. The schematic of the sample preparation device: 1—gas exhaust; 2—furnace door; 3—muffle furnace; 4—nitrogen supply; 5—volumetric flow meter; 6—reducing valve; 7—nitrogen supply tank; 8—samples.

The samples prepared were subsequently characterised by their proximate and ultimate analyses. Volatile matter was determined according to EN ISO 18123 [25] and ash content was measured in compliance with EN ISO 18122 [26]. Fixed carbon was calculated according to:

$$FC = 100 - (VM + A) \tag{1}$$

where FC is fixed carbon content, VM is volatile matter content, and A is ash content.

Grounded and homogenized samples of torrefied wood were analysed (ultimate analysis) by the ELEMENTAR varioMACROcube instrument (Elementar Analysensysteme, Hanau, Alemanha). Ground and homogenization of samples were performed by Grindomix GM 200 knife mill (Retsch GmbH, Haan, Germany) at speed 10,000 min^{-1} during 10 s.

The higher heating values of the samples were measured by the IKA C4000 (IKA Analysentechnik, Heitersheim, Germany) adiabatic calorimeter.

An important indicator of torrefaction is the energy yield, which indicates how much energy remains in the samples. Applying the relationship indicated in the work of Bach and Skrieberg, the energy yield of torrefied wood may be calculated as follows [27]:

$$Y_E = \frac{m_{torrefied}}{m_{raw}} \times \frac{HHV_{torrefied}}{HHV_{raw}} \times 100\% \tag{2}$$

where $m_{torrefied}$ is the mass of torrefied wood (kg), m_{raw} is the mass of raw wood (kg), $HHV_{torrefied}$ is the higher heating value of torrefied wood (MJ/kg) and $HHV_{torrefied}$ is the higher heating value of raw wood (MJ/kg).

The measurements were carried out using a cone calorimeter (Figure 2) according to ISO 5660-1 [28]. The sample (2) was covered by aluminium foil on the surfaces that had not been exposed to the heat flux and were inserted into the holder (1). The holder was subsequently placed underneath the cone heater (4). The combustion gases were exhausted via an exhaust hood (5), with the rate of the exhaust of the thermal decomposition products regulated by adjustment of the fan (7). The extraction tube contained a circular perforated probe (6), through which the combustion gases were sampled and analysed in the CO, CO_2 and O_2 analysers (8).

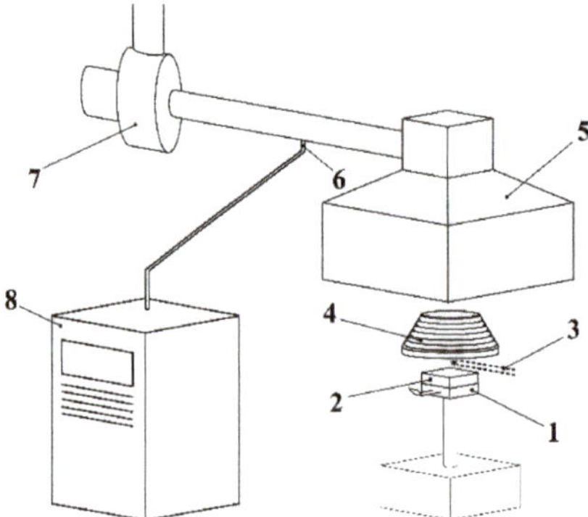

Figure 2. The schematic of the cone calorimeter: 1—sample holder, 2—sample, 3—initiator, 4—cone heater, 5—exhaust hood, 6—combustion gas sample extraction, 7—fan, 8—combustion gas analyser.

The fan flow rate was set to 0.024 ± 0.002 m^3 s^{-1}, ambient temperature ranged from 22 °C to 27 °C and the relative humidity of air was 20–27%. The atmospheric pressure was between 100.92–101.91 kPa. Measurements were performed at heat fluxes of 20 kW m^{-2}, 30 kW m^{-2} and 40 kW m^{-2}. Sampling interval was set to 5 s and grinding time was 1800 s.

Fuel quality is expressed by the combustion efficiency. According to Ferek et al. the combustion efficiency of biomass can be calculated by the Equation (3) [29]:

$$CE = \frac{\frac{[C]_{CO2}}{([C]_{CO2}+[C]_{CO})} - 0.18}{0.82} \tag{3}$$

where CE is the combustion efficiency, $[C]_{CO2}$ is the carbon emitted as CO_2, and $[C]_{CO}$ is the carbon emitted as CO.

The relationship characterising the time necessary for initiation can be written as [30]:

$$t_i = \frac{\pi}{4} k\rho c \left(\frac{T_i - T_0}{q_e}\right)^2 \tag{4}$$

where t_i is the time to ignition, k is thermal conductivity, ρ is density, c is heat capacity, T_i is ignition temperature, T_0 is ambient temperature, and q_e is external heat flux. This equation may be adjusted as follows:

$$TRP = (T_i - T_0)k\rho c \tag{5}$$

$$\sqrt{\frac{1}{t_i}} = \frac{2}{\sqrt{\pi}} \frac{q_e}{TRP} \tag{6}$$

where TRP is the thermal response parameter. According to Xu et al., TRP is used as an indicator of the ignition resistance of a material [31].

Hence, the critical heat flux (q_{cr}) is calculated as [32]:

$$\frac{q_i}{q_{cr}} = 0.76 \tag{7}$$

where q_i is the external heat flux with an infinite time necessary for initiation.

Therefore, the formula $\sqrt{\frac{1}{t_i}}$ of q_e allows the identification of the value of critical heat flux and the thermal response parameter. An advantage of the TRP calculation by this method is that no data concerning density, heat capacity and thermal conductivity at moment of ignition are needed.

Using the Stefan-Boltzman law, it is possible to conclude that:

$$T_{ig} = \sqrt[4]{\left(\frac{\alpha q_{cr}}{\sigma} + T_\infty{}^4\right)} \tag{8}$$

where T_{ig} is the ignition temperature at the critical heat flux and α is absorptivity, which is equal to emissivity.

Impact of wood species on the average HRR and combustion efficiency was evaluated by the Analysis of Variance (ANOVA) at a significance level $\alpha = 0.05$. Wood species with statistically equal combustion efficiency were revealed by the Duncan's test. The StatSoft STATISTICA 10 software was used for the ANOVA and Duncan's test.

3. Results and Discussion

The mass of raw and torrefied samples are given in Table 1. The yield of torrefied wood represented 37.77–46.40%. The lowest value corresponds to ash and the highest to oak.

Table 1. The mass of samples before and after the torrefaction.

Torrefied Wood	Weight of Raw Samples (g)	Weight of Torrefied Samples (g)	Mass Yield of Torrefied Wood (%)
Ash	127.95 (±2.01)	48.31 (±2.39)	37.77 (±2.13)
Beech	131.11 (±4.28)	50.52 (±3.24)	38.51 (±1.65)
Oak	127.06 (±3.12)	58.91 (±3.30)	46.40 (±3.05)
Pine	88.28 (±3.04)	38.98 (±1.44)	44.20 (±2.30)
Spruce	81.05 (±1.88)	35.91 (±1.33)	44.30 (±1.23)

Numbers in parentheses represent standard deviation.

Proximate and ultimate analysis of individual torrefied wood samples (Table 2) indicates a high carbon content, largely in the form of fixed carbon. Volatile matter represent 36.78–44.66% and ash ranges from 0.44% to 1.11%. In terms of elemental composition the amount of hydrogen appears to be relatively low. Athough, when converted to the amount of substance, it exceeds the oxygen content. Nitrogen and sulfur were present in the samples in negligible amounts.

Table 2. Proximate and ultimate analysis of torrefied wood samples.

Torrefied Wood	Proximate Analysis (%)			Ultimate Analysis (%)				
	Volatile Matter	Fixed Carbon	Ash	C	H	O [a]	N	S
Ash	36.78 (±2.68)	62.11 (±2.54)	1.11 (±0.31)	71.27 (±0.62)	3.79 (±0.07)	23.52 (±0.22)	0.22 (±0.04)	0.08 (±0.03)
Beech	41.88 (±0.34)	57.16 (±0.18)	0.96 (±0.24)	70.55 (±0.57)	4.03 (±0.07)	24.23 (±0.27)	0.23 (±0.07)	0.00 (±0.00)
Oak	42.16 (±0.35)	57.41 (±0.53)	0.44 (±0.20)	69.57 (±0.56)	4.21 (±0.09)	25.61 (±0.28)	0.16 (±0.06)	0.01 (±0.01)
Pine	44.60 (±1.67)	54.70 (±1.69)	0.70 (±0.07)	72.57 (±0.72)	4.61 (±0.08)	21.91 (±0.30)	0.19 (±0.03)	0.02 (±0.01)
Spruce	42.53 (±2.47)	56.80 (±2.26)	0.67 (±0.22)	70.59 (±0.63)	4.27 (±0.06)	24.34 (±0.25)	0.13 (±0.03)	0.00 (±0.00)

Numbers in parentheses represent standard deviation; [a] By calculation.

The high heating value was very similar in all torrefied wood samples (Table 3). Pine was slightly different from other types of wood. Energy yield ranged between approximately 54.5% and 66.5%. The ratio of O/C and H/C was 0.23–0.28 and 0.63–0.76, respectively.

The cone calorimeter was used to measure the time of initiation of combustion and heat release rate (Figure 3, Table 4), as well as the overall amount of carbon oxides released, which were used to calculate the combustion efficiency for each sample (Table 5).

Table 3. The energy characteristics of samples of torrefied wood.

Torrefied Wood	HHV (kJ g^{-1})	Energy Yield (%)	Atomic O/C Ratio (-)	Atomic H/C Ratio (-)
Ash	28.79 ($\pm$0.3)	54.57 ($\pm$2.90)	0.26	0.63
Beech	28.25 ($\pm$0.7)	55.02 ($\pm$1.99)	0.27	0.68
Oak	28.18 ($\pm$0.1)	66.58 ($\pm$4.45)	0.28	0.72
Pine	30.26 ($\pm$1.0)	63.83 ($\pm$1.09)	0.23	0.76
Spruce	28.64 ($\pm$0.7)	63.22 ($\pm$1.94)	0.27	0.72

Numbers in parentheses represent standard deviation.

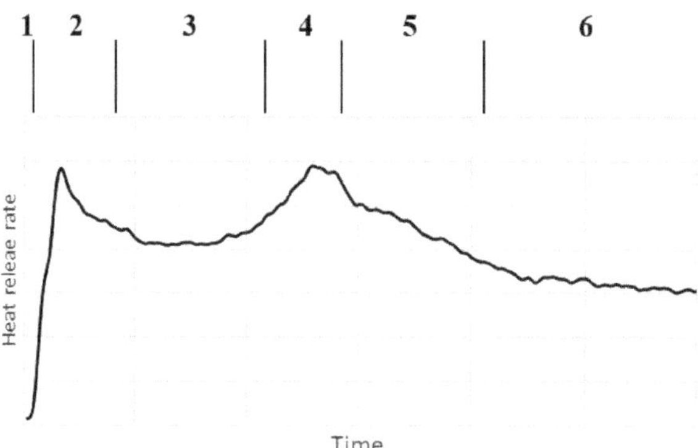

Figure 3. Phases of thermal degradation of torrefied wood: 1—pre-initiation phase, 2—initiation phase, 3—even combustion phase, 4—sample overheat phase, 5—low combustion phase, 6—heterogeneous combustion phase.

Table 4. Cone calorimeter results.

Torrefied Wood	External Heat Flux	pHRR [kW m^{-2}] First	pHRR [kW m^{-2}] Second	Time to pHRR [s] First	Time to pHRR [s] Second	Time to Ignition [s]	Average HRR [kW m^{-2}] 300 s	Average HRR [kW m^{-2}] 600 s	Average HRR [kW m^{-2}] 1200 s
Ash	20	27.08	-	240	-	180	23.59	23.62	21.57
	30	44.72	33.44	105	260	69	30.74	29.71	27.05
	40	57.68	60.00	55	535	35	44.92	48.37	42.43
Beech	20	47.52	30.83	275	710	246	28.67	28.18	26.53
	30	56.49	71.09	125	615	102	38.43	48.17	42.33
	40	57.79	91.54	70	515	51	46.56	54.31	46.76
Oak	20	22.82	31.74	285	1745	165	19.23	20.03	22.61
	30	56.22	29.86	130	655	107	32.79	29.72	28.08
	40	72.78	54.34	40	670	24	43.68	42.92	42.90
Pine	20	39.75	28.19	215	490	181	27.26	26.56	25.01
	30	86.86	64.12	60	605	45	47.24	50.37	42.60
	40	83.620	91.99	30	560	17	58.34	67.01	51.61
Spruce	20	44.16	34.38	130	1695	110	28.05	27.59	28.75
	30	63.43	-	50	-	33	35.07	30.76	26.73
	40	83.99	49.08	30	125	17	46.79	44.25	35.12

Table 5. Results of two-way ANOVA examining both the impact of wood species and heat flux on the average HRR during investigated time intervals (300, 600 and 1200 s) at a significance level $\alpha = 0.05$.

ANOVA Coefficients	Average HRR in Time Interval [s]		
	300	600	1200
p (wood species)	0.0108	0.0218	0.1009
F (wood species)	6.8152	5.3207	2.7934
F_{crit} (wood species)	3.8379	3.8379	3.8379
p (heat flux)	1.38×10^{-5}	0.0002	0.0009
F (heat flux)	61.62	28.65	18.65
F_{crit} (heat flux)	4.4589	4.4589	4.4589

As to the visual comparison and rate of heat release, the process indicated in the individual charts may be divided into 6 phases.

1. The pre-initiation phase (the rate of heat release is essentially equal to zero, no visual changes in the samples can be observed),
2. The initiation phase (heat release rate rapidly increases and then falls, it is possible to see the beginning of combustion),
3. The even combustion phase (the rate of heat release is relatively constant, combustion appears even),
4. The sample overheat phase (the heat release rate increases and reaches its second peak, it is possible to observe a stronger flame),
5. The low combustion phase (the heat release rate decreases, it is possible to observe a decrease in the intensity of the flame, leading to extinction), and
6. The heterogeneous combustion phase (the speed of heat release slowly decreases; it is possible to observe blazing of the sample).

The phases are shown in Figure 3. For investigated samples of torrefied wood these phases can be recognized in Figure 4.

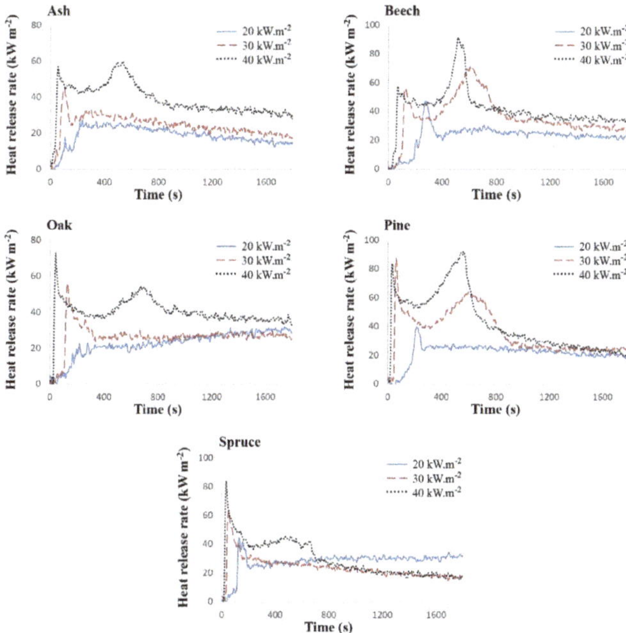

Figure 4. The change in heat release rate over time during the cone calorimeter measurements.

Heat release rate curve with two peaks is common for thermally thick materials. The first peak corresponds to the combustion of volatile combustibles before the formation of the carbonized layer [33,34] and the second is very sensitive to the thickness of the insulating substrate [35].

Impact of external heat flux and wood species on phases of thermal degradation is different. Increase of the heat flux causes greater heating of samples. This greater heating results in more clearly distinguished phases. All phases can be distinguished for all investigated samples at heat flux of 40 kW·m^{-2}. On the other hand, oak, spruce and ash samples have more pronounced pre-initiation and initiation phases, while other phases are not sharply distinguished under the heat flux of 30 kW·m^{-2}. Under the heat flux of 20 kW·m^{-2} only the first phase and second phase can be seen in all cases.

Both the time to ignition and thus also the time duration of pre-initiation phase of torrefied wood are dependent mainly on the external heat flux (Figure 5). The heat flux radiated to the surface of the sample results in heating of the top layer of material. Heating the material to a higher temperature results in a faster release of flammable degradation products. When mixed with an oxidizing agent (mostly atmospheric oxygen), a flammable composition capable of initiation is formed. This phenomenon is well known and commonly used for ignition parameters calculation.

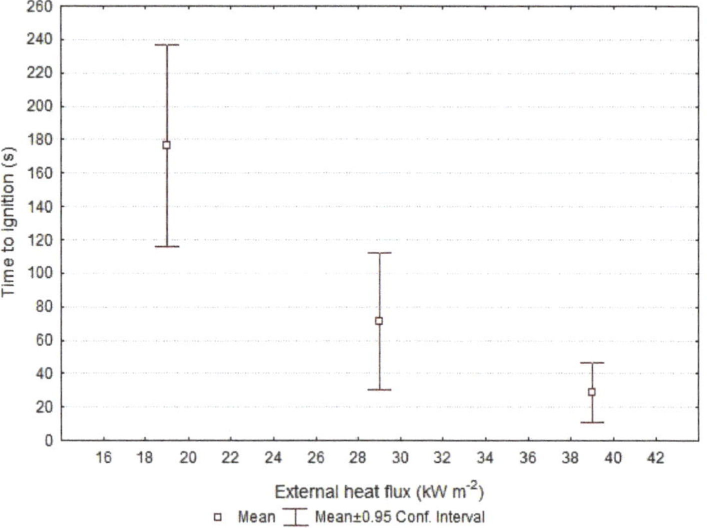

Figure 5. Impact of external heat flux on the time to ignition of torrefied wood.

Time duration of initiation phase was in the range from 51 to 104 s for all samples. Impact of external heat flux on the time duration of this phase was not statistically significant.

The results clearly indicate that the higher the external heat flux, the higher the combustion rate. The heat release rate values are also higher and their peaks are shifted towards the beginning of the test.

If the external heat flux is 40 kW m^{-2}, almost all the samples have two peaks. With an external heat flux of 30 kW m^{-2}, the sample overheat phase is less significant, and in the case of spruce it is practically non-existent. When the samples were exposed to an external heat flux of 20 kW m^{-2}, the second peak was negligible.

In general, in torrefied ash and torrefied oak, the values of released heat are almost identical to the amount of heat to which the surface of the samples is exposed. On the contrary, the highest average heat release rates were achieved by torrefied pine and torrefied beech.

ANOVA (two-way ANOVA at a significance level of $\alpha = 0.05$) results revealing both impact of wood species and heat flux on the average heat release rate for three time intervals

(300, 600 and 1200 s) are in the Table 5. The data in Table 5 proved that impact of wood species on the average heat release rate is only statistically significant for 300 and 600 s time interval. Moreover, Table 5 proved statistically significant impact of heat flux on average heat release rate for all investigated time intervals (300, 600 and 1200 s). The lowest value of the heat release rate was reached by torrefied oak. Since this type of wood has high resistance to ignition and burning even in the untreated state [36], it can be assumed that it retains similar properties compared to other woods even after the torrefaction process.

The calculated combustion efficiency values are listed in Table 6. As the external heat flux increases, so the combustion efficiency also increases, reaching, on average, less than 71% at 20 kW m^{-2}, more than 81% at 30 kW m^{-2} and almost 90% at 40 kW m^{-2}. The reason for the increase in combustion efficiency with increasing external heat flux is that at higher heat flux levels there is more pronounced oxidation of the solid carbonaceous layer formed on the sample during the cone calorimeter test.

Table 6. Combustion efficiencies of torrefied wood obtained using different heating fluxes.

Torrefied Wood	Combustion Efficiency [%]		
	$q_e = 20$ kW m^{-2}	$q_e = 30$ kW m^{-2}	$q_e = 40$ kW m^{-2}
Ash	75.1 (6.8) [bc]	80.7 (9.5)	89.4 (5.4)
Beech	74.7 (7.2) [abd]	87.9 (5.8) [a]	92.1 (4.5) [a]
Oak	61.8 (10.2)	73.1 (9.3)	88.5 (5.5)
Pine	72.3 (6.9) [a]	87.6 (6.1) [a]	92.2 (4.2) [a]
Spruce	74.4 (7.0) [cd]	77.8 (4.8)	85.7 (7.4)

Numbers in parentheses represent standard deviation, [a] wood species with statistically equal combustion efficiencies at all investigated heat fluxes, [b,c,d] wood species with statistically equal combustion efficiency at heat flux of 20 kW·m^{-2}.

ANOVA results of the impact of wood species on the combustion efficiency for investigated heat fluxes of 20, 30 and 40 kW·m^{-2} are in the Table 7. Data in the Table 7 proved that the type of wood species has statistically significant impact on the combustion efficiency.

Table 7. Results of ANOVA examining the impact of wood species on the combustion efficiency for heat fluxes of 20, 30 and 40 kW·m^{-2} (at significance level $\alpha = 0.05$).

ANOVA Coefficients	Heat Flux [kW·m^{-2}]		
	20	30	40
p	3.9×10^{-137}	2.7×10^{-183}	1×10^{-66}
F	192.03	272.77	85.75
F_{crit}	2.38	2.38	2.38

ANOVA is able to evaluate if there are statistically significant differences between investigated samples. However, this method is not able to evaluate between which samples are significant differences. The Duncan's test was used for this purpose. The results of the Duncan's test are implemented to Table 6. The obtained results proved that in all investigated heat fluxes (from 20 to 40 kW·m^{-2}), the difference between the pine wood and the beech wood combustion efficiency are not significant (Duncan's test p value is higher than 0.05). At heat flux of 20 kW·m^{-2}, the differences between the beech and ash wood, between spruce and ash wood and between spruce and beech wood are statistically insignificant.

By simplifying the situation and stating that the surface of torrefied wood behaves like a black body, the emissivity of torrefied wood becomes 1. The initiation temperatures calculated in this way, as well as the critical heat fluxes, the thermal response parameters and the respective determination coefficients, are indicated in Table 8.

Table 8. Critical heat flux and thermal response parameter of torrefied wood.

Torrefied Wood	Critical Heat Flux [kW m^{-2}]	Thermal Response Parameter [kW s$^{0.5}$ m^{-2}]	R^2 [-]	Ignition Temperature [°C]
Ash	5.7	240	0.9997	303
Beech	4.67	297	0.9981	277
Oak	13.2	179	0.8590	428
Pine	15.2	134	0.9959	452
Spruce	8.9	152	0.9984	365

For solution of many tasks regarding fire safety of polymers average values of ignition parameters are very important. The average values of the most important ignition parameters of torrefied wood are in the Table 9.

Table 9. Average ignition parameters of torrefied wood.

Ignition Parameter	Value ± Standard Deviation
Critical heat flux [kW·m^{-2}]	9.5 ± 4.6
Thermal response parameter [kW·s$^{0.5}$·m^{-2}]	200.4 ± 67.3
Ignition temperature [°C]	365 ± 76

The yield of torrefied wood decreases with increasing temperature and time. For pine, Burgois and Guyonnet state that after 4 h at a temperature of 260 °C, it fell to 50.13%. It contained 70.71% of carbon and 24.49% of oxygen and 4.66% of hydrogen. The volatile combustible matter was 47.6% [37]. These values resemble the data that characterises the prepared torrefied wood samples. Although cited authors prepared torrefied wood at lower temperature, its influence was compensated by the longer time interval.

At 290 °C, Manouchehrinejad, van Giesen and Mani report a significantly higher volatile matter content (63.57) and a lower amount of fixed carbon (35.62) [38]. However, in the torrefaction process they used, the wood chips were exposed to an increased temperature for only 30 min. For the case of wood pellets of the torrefied wood mentioned above, the measured components are slightly closer to those of our samples.

Lee et al. also indicate that the ratio of volatile matter/fixed carbon. They report a value of 0.78 for torrefied wood pellets prepared at a temperature of 300 °C for at least 4 h, which corresponds to the values from our measurements (0.59–0.81). The carbon content (74.8%) and higher heating value (28.8 kJ g^{-1}) are also similar. The hydrogen content is higher (5.1%) and the oxygen content is lower (19.2%). The energy yield is also slightly higher (69.6%) [39].

Strandberg et al. prepared torrefied wood from spruce at temperature of 310 °C during 25 min. The mass yield in the above-mentioned study (46%) was higher than mass yield from spruce prepared in this work. On the other hand energy yield published by Strandberg et al. was slightly lower (62%) than energy yield of spruce wood in this study. The elemental composition of torrified spruce wood in both studies were very similar (sample in this study contained slightly more carbon and less hydrogen and oxygen). Significant difference between torrified spruce wood was in volatile matter (51.5%) and fixed carbon (47.8%) stated in this and above-mentioned study [40]. The obtained results proved slightly higher degree of spruce wood torrefaction caused by longer duration of heat load.

Energy yield of pine wood sawdust torrefied at 300 °C for 6 min is 85.71% with higher heating value of 22.35 MJ kg^{-1} [41]. Similar to [41] the degree of torrefaction is much lower than in the case of torrefied pine at 300 ° C for 120 min due to the short exposure time of wood to high temperature.

Magdiarz, Wilk and Straka prepared (by torrefaction of fuel wood at temperature of 290 °C during 60 min) product that contains: 62.5–66.4% of carbon and 4.48–4.56% of hydrogen. Calorific value of this product was 24.4 MJ kg^{-1}–26.2 MJ kg^{-1}. Mass yield and energy yield were 39–43%, and 58–61%, respectively [13]. These values are almost the same as values obtained in this study. Although, the cited authors used a shorter time period in thermal loading, they prepared very similar product (the cause was the use of lower sized samples in the cited paper).

Solid fuels are always characterised based on their elementary H/C/O balances. A Van Krevelen diagram shows that there is a clear increase in the heating value of the different solid fuels by increasing the H/C and decreasing the O/C ratios [42]. The ranking of the results of the torrified wood samples compared to other fuels is shown in Figure 6. Torrefied pine clearly has similar features to coal. Similarly, Elaieb et al. described the charcoal produced by carbonization at a temperature of 550 °C over the course of 6 h [43].

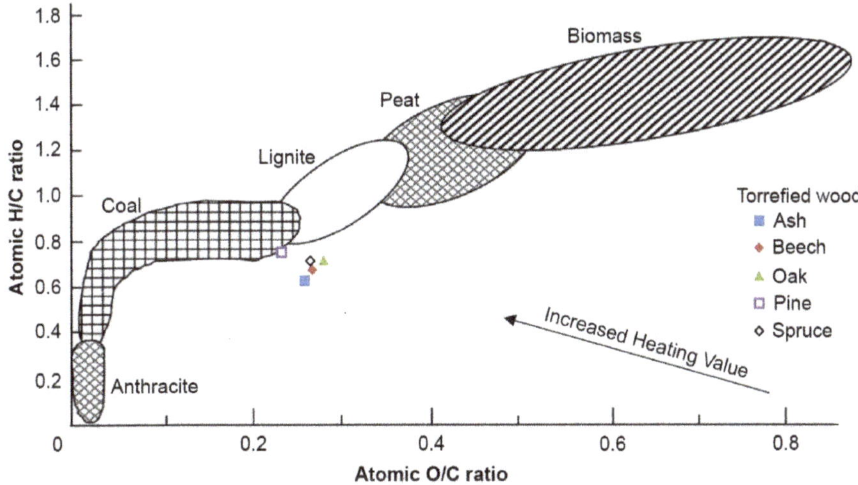

Figure 6. Graphic representation of the tested samples in a Van Krevelen diagram (based on Prins et al. [44]).

As to safe storage, it is necessary to evaluate the ability of the individual materials to contribute to the ignition and spread of fire. It is important to know their reactions to sources of radiant heat, which include both hot surfaces (e.g., heaters) and flame re-radiation. The cone calorimeter measurements were used for this purpose. As mentioned above, there were two peaks in the measurement of the heat release rate. It is well-known that the first peak is linked to the combustion ignition. The second one was recorded at the end of the measurements. This process is also typical of untreated wood. When a sample of finite thickness is burned in a heat release calorimeter, the HRR increases toward the end of the test as a result of the near adiabatic conditions on the unexposed side [45]. The effective heat of pyrolysis is low when the thermal wave reaches the rear insulating surface and the original material is already preheated to the pyrolysis temperature [33].

Several authors have observed the effect of thermal treatment of the wood on the rate of heat release during combustion. Luptakova et al. states that heat treatment of wood at temperatures of 200–260 °C resulted in a lower mass loss a lower average relative burning rate, but it did not influence ignition time, the flame-out time, and maximum burning rate [46]. Based on measurements taken at external heat fluxes of 15–40 kW m^{-2} Martinka et al. state that the heat treatment of spruce causes a significant decrease in the maximum heat release rate [47]. Xing and Li. reached similar conclusions [48]. Lahtela and Kärki impregnated thermally treated wood with melamine and found that the heat treatment reduced the HRR values, but melamine impregnation before heat treatment was able to raise it to a higher value [49]. The aforementioned values of the peak heat release rates are significantly higher than for torrefied wood, which may be ascribed to the significantly lower temperatures used for the thermal treatment (180–220 °C) as opposed those used in torrefaction.

Elaieb et al. used a cone calorimeter to directly test the carbonized wood. However, they employed an oil burner as an initiator and observed the ignition time, combustion duration; combustion states and smoke [43]. For these reasons, it is impossible to compare the sets of results.

The critical heat fluxes calculated based on the initiation times, the thermal response parameters and initiation temperatures of torrefied wood resemble those stated by other authors for different types of wood (Table 10). Hence, the samples may be classified into two groups based on the calculated values of critical heat flux: Torrefied ash, torrefied beech and torrefied spruce reach values of less than 10 kW m^{-2}, and torrefied oak and torrefied pine over 10 kW m^{-2}. Nonetheless, for torrefied oak, the correlation coefficient of the corresponding equation is significantly lower, which is why the values of critical heat flux, heat response parameter and initiation temperature are only indicative.

Table 10. Critical heat fluxes and thermal response parameters of selected types of wood.

Wood	Critical Heat Flux [kW m^{-2}]	Thermal Response Parameter [kW s$^{0.5}$ m^{-2}]	Ignition Temperature [°C]	Source
Douglas fir	18	182	478	[31]
Scots pine	19	164	488	[31]
Southern pine	19	201	488	[31]
Shorea	16	152	456	[31]
Merbau	40	275	643	[31]
Redwood	15.5	-	375	[32]
Red oak	108	-	304	[32]
Douglas fir	16.0	-	384	[32]
Maple	13.9	-	354	[32]
Nordic spruce	19.0	291	488	[33]
Fir	11.6–12.0	128–144	372.7	[50]
Radiata pine	13.2	-	-	[51]
Pacific maple	10.3	-	-	[51]
Sugar pine	14.0	-	-	[51]
Bamboo	6.0–8.0	235–376	297–340	[52]
Spruce	10.1	-	-	[53]
Softwood	10.0	-	-	[54]
Leadwood	15.0	376.2	-	[55]
Mopani	14.4	161.2	-	[55]
Tamboti	5.9	352.7	-	[55]
Stinkwood	9.2	173.6	-	[55]
Real Yellowwood	1.3	232.2	-	[55]

4. Conclusions

Based on the measurements conducted on samples of torrefied wood from five different types of wood, it was discovered that the placement of such fuel in the van Krevelen chart is close to coal and lignite. Treatment at 300 °C for 2 h under nitrogen also appears to be sufficient for samples with dimensions of 100 mm × 100 mm × 20 mm with an energy yield from 49.45 to 61.09%. The samples were measured on cone calorimeter in a compact form. Therefore, the obtained results are suitable for use especially in places where torrefied wood does not occur in the crushed state.

The heat release rate increases with increasing external heat flux, although it also increases unsteadiness of combustion. Two clear peaks occur in the heat release rate at an external heat flux of 40 kW m^{-2}, but these are significantly lower than those from the thermally untreated biomass. The combustion of the torrefied wood while making measurements using a cone calorimeter can be divided into 6 phases: Pre-initiation phase, the initiation phase, the even combustion phase, the sample overheat phase, the low combustion phase and the heterogeneous combustion phase.

The combustion efficiency identified based on the amount of CO and CO_2 in the combustion gases increases as the external heat flux increases. On average, it reaches almost 71% at a heat flux of 20 kW m^{-2}, more than 81% at 30 kW m^{-2} and almost 91% at 40 kW m^{-2}.

Torrefied wood increases the fire load of fire compartments during storage (in comparison with unmodified wood). The obtained results are key for designing the fire safety of buildings where this material is stored.

Author Contributions: Conceptualization, P.R.; methodology, P.R. and A.H.; formal analysis, P.R. and J.M.; investigation, P.R. and A.H.; resources, J.M.; data curation, P.R., J.M. and A.H.; writing—original draft preparation, P.R.; writing—review and editing, P.R. and J.M.; visualization, P.R.; supervision, J.M.; project administration, P.R.; funding acquisition, J.M. All authors have read and agreed to the published version of the manuscript.

Funding: This work was supported by the Slovak Research and Development Agency under the contract No. APVV-16-0223. This work was also supported by the KEGA Agency under the contract No. 016STU-4/2021. This work was supported by the Scientific Grant Agency, project VEGA 1/0403/19.

Conflicts of Interest: The authors declare no conflict of interest. The funders had no role in the design of the study; in the collection, analyses, or interpretation of data; in the writing of the manuscript, or in the decision to publish the results.

References

1. Zorpas, A.A.; Pociovălişteanu, D.M.; Georgiadou, L.; Voukkali, I. Environmental and technical evaluation of the use of alternative fuels through multi-criteria analysis model. *Prog. Ind. Ecol.* **2016**, *10*, 3–15. [CrossRef]
2. Brundtland Commission. Our common future, Chapter 2: Towards sustainable development. In *World Commission on Environment and Development (WCED)*; United Nation: Geneva, Switzerland, 1987.
3. Medic, D.; Darr, M.; Potter, B.; Shah, A. Effects of torrefaction process parameters on biomass feedstock upgrading. *Fuel* **2012**, *91*, 147–154. [CrossRef]
4. International Energy Agency. *Renewables Information: Overview*; International Energy Agency: Paris, France, 2017.
5. Lu, K.M.; Lee, W.J.; Chen, W.H.; Lin, T.C. Thermogravimetric analysis and kinetics of co-pyrolysis of raw/torrefied wood and coal blends. *Appl. Ener.* **2013**, *105*, 57–65. [CrossRef]
6. Chen, W.H.; Kuo, P.C. Isothermal torrefaction kinetics of hemicellulose, cellulose, lignin and xylan using thermogravimetric analysis. *Energy* **2011**, *36*, 6451–6460. [CrossRef]
7. Van der Stelt, M.J.C.; Gerhauser, H.; Kiel, J.H.A.; Ptasinski, K.J. Biomass upgrading by torrefaction for the production of biofuels: A review. *Biomass Bioenergy* **2011**, *35*, 3748–3762. [CrossRef]
8. Bridgeman, T.G.; Jones, J.M.; Shield, I.; Williams, P.T. Torrefaction of reed canary grass, wheat straw and willow to enhance solid fuel qualities and combustion properties. *Fuel* **2008**, *87*, 844–856. [CrossRef]
9. Prins, M.J.; Ptasinski, K.J.; Janssen, F.J.J.G. Torrefaction of wood: Part 1. Weight loss kinetics. *J. Anal. Appl. Pyrolysis* **2006**, *77*, 28–34. [CrossRef]
10. Bourgois, J.; Bartholin, M.C.; Guyonnet, R. Thermal treatment of wood: Analysis of the obtained product. *Wood Sci. Technol.* **1989**, *23*, 303–310. [CrossRef]
11. Chen, W.H.; Hsu, H.C.; Lu, K.M.; Lee, W.J.; Lin, T.C. Thermal pretreatment of wood (Lauan) block by torrefaction and its influence on the properties of the biomass. *Energy* **2011**, *36*, 3012–3021. [CrossRef]
12. Wannapeera, J.; Worasuwannarak, N. Examinations of chemical properties and pyrolysis behaviors of torrefied woody biomass prepared at the same torrefaction mass yields. *J. Anal. Appl. Pyrol.* **2015**, *115*, 279–287. [CrossRef]
13. Magdziarz, A.; Wilk, M.; Straka, R. Combustion process of torrefied wood biomass. *J. Therm. Anal. Calorim.* **2017**, *127*, 1339–1349. [CrossRef]

14. Mi, B.; Liu, Z.; Hu, W.; Wei, P.; Jiang, Z.; Fei, B. Investigating pyrolysis and combustion characteristics of torrefied bamboo, torrefied wood and their blends. *Biores. Technol.* **2016**, *209*, 50–55. [CrossRef] [PubMed]

15. Phanphanich, M.; Mani, S. Impact of torrefaction on the grindability and fuel characteristics of forest biomass. *Bioresour. Technol.* **2011**, *102*, 1246–1253. [CrossRef] [PubMed]

16. Prins, M.J.; Ptasinski, K.J.; Janssen, F.J. More efficient biomass gasification via torrefaction. *Energy* **2006**, *31*, 3458–3470. [CrossRef]

17. Statistical Office of the Slovak Republic. *Waste in Slovak Republic in 2019*; Statistical Office of the Slovak Republic: Bratislava, Slovakia, 2020; p. 99.

18. Antov, P.; Savov, V.; Krišťák, L.; Réh, R.; Mantanis, G.I. Eco-friendly, high-density fiberboards bonded with urea-formaldehyde and ammonium lignosulfonate. *Polymers* **2021**, *13*, 220. [CrossRef]

19. Statistical Office of the Slovak Republic. *Forest Management in the Slovak Republic in 2005–2009*; Statistical Office of the Slovak Republic: Bratislava, Slovakia, 2010; p. 53.

20. Ministry of Agriculture and Development of the Slovak Republic, National Forest Centre. Green report. In *Report on the Forest Sector of the Slovak Republic 2019*; Ministry of Agriculture and Development of the Slovak Republic, National Forest Centre: Bratislava, Slovakia, 2020; p. 68.

21. Vandličková, M.; Marková, I.; Makovická Osvaldová, L.; Gašpercová, S.; Svetlík, J.; Vraniak, J. Tropical wood dusts—granulometry, morfology and ignition temperature. *Appl. Sci.* **2020**, *10*, 7608. [CrossRef]

22. Wang, Z.; Ning, X.; Zhu, K.; Hu, J.; Yang, H.; Wang, J. Evaluating the thermal failure risk of large-format lithium-ion batteries using a cone calorimeter. *J. Fire Sci.* **2019**, *37*, 81–95. [CrossRef]

23. Cai, W.; Zhao, Z.; Wang, W.; Guo, W.; Wang, X.; Hu, Y. Combustion behavior characterization of major crops through cone calorimeter. *Fire Mater.* **2020**, *44*, 693–703. [CrossRef]

24. Liu, Z.; Hu, W.; Jiang, Z.; Mi, B.; Fei, B. Investigating combustion behaviors of bamboo, torrefied bamboo, coal and their respective blends by thermogravimetric analysis. *Renew. Energ.* **2016**, *87*, 346–352. [CrossRef]

25. EN ISO 18123:2015—Solid biofuels. *Determination of the Content of Volatile Matter*; International Organization for Standardization: Geneva, Switzerland, 2015.

26. EN ISO 18122: 2015—Solid Biofuels. *Determination of Ash Content*; International Organization for Standardization: Geneva, Switzerland, 2015.

27. Bach, Q.V.; Skreiberg, Ø. Upgrading biomass fuels via wet torrefaction: A review and comparison with dry torrefaction. *Renew. Sustain. Energy Rev.* **2016**, *54*, 665–677. [CrossRef]

28. ISO 5660-1:2015—Reaction to fire tests. *Heat Release, Smoke Production and Mass Loss Rate. Part 1: Heat Release Rate (Cone Calorimeter Method) and Smoke Production Rate (Dynamic Measurement)*; International Organization for Standardization: Geneva, Switzerland, 2015.

29. Ferek, R.J.; Reid, J.S.; Hobbs, P.V.; Blake, D.R.; Liousse, C. Emission factors of hydrocarbons, halocarbons, trace gases and particles from biomass burning in Brazil. *J. Geophys. Res. Atmos.* **1998**, *103*, 32107–32118. [CrossRef]

30. Fateh, T.; Rogaume, T.; Luche, J.; Richard, F.; Jabouille, F. Characterization of the thermal decomposition of two kinds of plywood with a cone calorimeter—FTIR apparatus. *J. Anal. Appl. Pyrol.* **2014**, *107*, 87–100. [CrossRef]

31. Xu, Q.; Chen, L.; Harries, K.A.; Zhang, F.; Liu, Q.; Feng, J. Combustion and charring properties of five common constructional wood species from cone calorimeter tests. *Constr. Build. Mater.* **2015**, *96*, 416–427. [CrossRef]

32. Spearpoint, M.J.; Quintiere, J.G. Predicting the piloted ignition of wood in the cone calorimeter using an integral model—effect of species, grain orientation and heat flux. *Fire Saf. J.* **2001**, *36*, 391–415. [CrossRef]

33. Hagen, M.; Hereid, J.; Delichatsios, M.A.; Zhang, J.; Bakirtzis, D. Flammability assessment of fire-retarded Nordic Spruce wood using thermogravimetric analyses and cone calorimetry. *Fire Saf. J.* **2009**, *44*, 1053–1066. [CrossRef]

34. Wang, Y.; Zhao, J. Facile preparation of slag or fly ash geopolymer composite coatings with flame resistance. *Constr. Build. Mater.* **2019**, *203*, 655–661. [CrossRef]

35. Ritchie, S.J.; Steckler, K.D.; Hamins, A.; Cleary, T.G.; Yang, J.C.; Kashiwagi, T. The effect of sample size on the heat release rate of charring materials. *Fire Saf. Sci.* **1997**, *5*, 177–188. [CrossRef]

36. Osvaldová Makovická, L.; Castellanos, J.-R.S. Burning rate of selected hardwood tree species. *Acta Fac. Xylologiae Zvolen Publica Slovaca* **2019**, *61*, 91–97.

37. Bourgois, J.; Guyonnet, R. Characterization and analysis of torrefied wood. *Wood Sci. Technol.* **1988**, *22*, 143–155. [CrossRef]

38. Manouchehrinejad, M.; Van Giesen, I.; Mani, S. Grindability of torrefied wood chips and wood pellets. *Fuel Process. Technol.* **2018**, *182*, 45–55. [CrossRef]

39. Lee, Y.; Yang, W.; Chae, T.; Kang, B.; Park, J.; Ryu, C. Comparative Characterization of a Torrefied Wood Pellet under Steam and Nitrogen Atmospheres. *Energy Fuel.* **2018**, *32*, 5109–5114. [CrossRef]

40. Strandberg, M.; Olofsson, I.; Pommer, L.; Wiklund-Lindström, S.; Åberg, K.; Nordin, A. Effects of temperature and residence time on continuous torrefaction of spruce wood. *Fuel Process. Technol.* **2015**, *134*, 387–398. [CrossRef]

41. Yang, I.; Cooke-Willis, M.; Song, B.; Hall, P. Densification of torrefied Pinus radiata sawdust as a solid biofuel: Effect of key variables on the durability and hydrophobicity of briquettes. *Fuel Process. Technol.* **2021**, *214*, 106719. [CrossRef]

42. Ranzi, E.; Faravelli, T.; Manenti, F. Pyrolysis, gasification, and combustion of solid fuels. *Adv. Chem. Eng.* **2016**, *49*, 1–94.

43. Elaieb, M.T.; Khouaja, A.; Khouja, M.L.; Valette, J.; Volle, G.; Candelier, K. Comparative study of local tunisian woods properties and the respective qualities of their charcoals produced by a new industrial eco-friendly carbonization process. *Waste Biomass Valori.* **2018**, *9*, 1199–1211. [CrossRef]

44. Prins, M.J.; Ptasinski, K.J.; Janssen, F.J.J.G. From coal to biomass gasification: Comparison of thermodynamic efficiency. *Energy* **2007**, *32*, 1248–1259. [CrossRef]
45. Tran, H.C.; White, R.H. Burning rate of solid wood measured in a heat release rate calorimeter. *Fire Mater.* **1992**, *16*, 197–206. [CrossRef]
46. Luptáková, J.; Kačík, F.; Mitterová, I.; Zachar, M. Influence of Temperature of Thermal Modification on the Fire-technical Characteristics of Spruce Wood. *BioResources* **2019**, *14*, 3795–3807.
47. Martinka, J.; Hroncová, E.; Chrebet, T.; Balog, K. The influence of spruce wood heat treatment on its thermal stability and burning process. *Eur. J. Wood Wood Prod.* **2014**, *72*, 477–486. [CrossRef]
48. Xing, D.; Li, J. Effects of heat treatment on thermal decomposition and combustion performance of *Larix* spp. wood. *BioResources* **2014**, *9*, 4274–4287. [CrossRef]
49. Lahtela, V.; Kärki, T. The influence of melamine impregnation and heat treatment on the fire performance of Scots pine (*Pinus sylvetris*) wood. *Fire Mater.* **2016**, *40*, 731–737. [CrossRef]
50. Batiot, B.; Luche, J.; Rogaume, T. Thermal and chemical analysis of flammability and combustibility of fir wood in cone calorimeter coupled to FTIR apparatus. *Fire Mater.* **2014**, *38*, 418–431. [CrossRef]
51. Moghtaderi, B.; Novozhilov, V.; Fletcher, D.F.; Kent, J.H. A new correlation for bench-scale piloted ignition data of wood. *Fire Saf. J.* **1997**, *29*, 41–59. [CrossRef]
52. Xu, Q.; Chen, L.; Harries, K.A.; Li, X. Combustion performance of engineered bamboo from cone calorimeter tests. *Eur. J. Wood Wood Prod.* **2017**, *75*, 161–173. [CrossRef]
53. Park, H.J.; Lee, S.M. Combustion Characteristics of Spruce Wood by Pressure Impregnation with Waterglass and Carbon Dioxide. *Fire Sci. Eng.* **2012**, *26*, 18–23. [CrossRef]
54. Shields, T.J.; Silcock, G.W.; Murray, J.J. Evaluating ignition data using the flux time product. *Fire Mater.* **1994**, *18*, 243–254. [CrossRef]
55. Maake, T.; Asante, J.; Mwakikunga, B. Fire performance properties of commonly used South African hardwood. *J. Fire Sci.* **2020**, *38*, 415–432. [CrossRef]

 *polymers*

Article

Experimental Study of Straw-Based Eco-Panel Using a Small Ignition Initiator

Linda Makovicka Osvaldova [1,*], Iveta Markova [1], Stanislav Jochim [2] and Jan Bares [3]

[1] Department of Fire Engineering, Faculty of Security Engineering, University of Zilina, Univerzitna 1, 010 26 Zilina, Slovakia; iveta.markova@fbi.uniza.sk
[2] Department of Wooden Constructions, Faculty of Wood Science and Technology, Technical University in Zvolen, T.G. Masaryka 24, 960 01 Zvolen, Slovakia; jochim@tuzvo.sk
[3] EKOPANELY SERVIS s.r.o., Jedousov 64, 535 01 Prelouc, Czech Republic; bares@ekopanely.cz
* Correspondence: linda.makovicka@fbi.uniza.sk

Abstract: Straw, a natural cellulose-based material, has become part of building elements. Eco-panels, compressed straw in a cardboard casing, is used as building insulation. Eco-panel is a secondary product with excellent insulating properties. If suitably fire-treated (insulation and covering), straw panels' fire resistance may be increased. This contribution deals with monitoring the behavior of eco-panels exposed to a small ignition initiator (flame). The samples consisted of compressed straw boards coated with a 40 mm thick cardboard. Samples were exposed to a flame for 5 and 10 min. The influence of the selected factors (size of the board, orientation of flame with the sample) were compared on the basis of experimentally obtained data: mass loss. The results obtained do not show a statistically significant influence of the position of the sample and the initiating source (flame). The results presented in the article confirm the justifiability of fire tests. As the results of the experiments prove, the position of a small burner for igniting such material is also important. Such weakness of the material can also be eliminated by design solutions in the construction. The experiment on larger samples also confirmed the justifiability of fire tests along with the need for flame retardancy of such material for its safe application in construction.

Keywords: eco-panel; small ignition initiator; straw; relative burning rate; weight loss; fire properties

Citation: Makovicka Osvaldova, L.; Markova, I.; Jochim, S.; Bares, J. Experimental Study of Straw-Based Eco-Panel Using a Small Ignition Initiator. *Polymers* **2021**, *13*, 1344. https://doi.org/10.3390/polym13081344

Academic Editor: Gianluca Tondi

Received: 23 March 2021
Accepted: 18 April 2021
Published: 20 April 2021

Publisher's Note: MDPI stays neutral with regard to jurisdictional claims in published maps and institutional affiliations.

1. Introduction

Eco-panel represents a natural material used in the construction industry [1,2]. Eco-panel is a fully recyclable material and, as a building element, is a secondary product made out of straw with a casing of recycled paperboard or cardboard [3]. Eco-panel is produced by pressing straw at 180–220 °C [4], using pressure (15 MPa), without any binders (pure straw core), and subsequently gluing it together on both sides with recycled cardboard or recycled paper [5,6]. A disadvantage of this material is that it is relatively heavy [7,8].

Eco-panel has excellent thermal insulation properties and low thermal conductivity [9,10]. The research on eco-panels has focused primarily on monitoring the quality of indoor climates in enclosed environments with such straw insulation [11], and a solution of low-energy wood houses insulated with eco-panel. Sadzevicius [12] determined, by standard methods, the main qualities of straw: heat conductivity, fire, and humidity resistance, durability, and strength of compressed straw. The compressive strength of straw panels, calculated by averaging, yielded a value of 144 ± 30 kPa [13]. They assert that there is no standard for compressive strength of panels, including straw [13].

The growing need for sustainable products and the stringent legislative requirements related to the hazardous formaldehyde emissions from wood-based panels have boosted scientific and industrial interest in the production of eco-friendly, wood-based panels and optimal utilization of the available lignocellulosic materials [14,15].

The fire characteristics of building materials play an important role in building design. Fire properties of natural materials are usually, in comparison with "standard" materials, worse [16].

The importance of correctly applying natural materials, such as straw, in constructions is illustrated by the fire in the construction of the Alternative Theater S2, built under an overpass in Zilina. The construction consisted of wooden bottle crates, straw bales (the inner lining made of a thousand straw bales), and clay [17]. The fire, which originated at 1:30 a.m. on 19 May 2019, left the structure completely destroyed. The fire spread quickly and the flames also damaged the construction of the overpass.

Fire Characteristics of Straw-Based Eco-Panel Constructions Coated with Cardboard

Fire characteristics of straw structures are a topic of discussion. Straw structures are said to be very susceptible to fire and burn very well [16,18].

There are separate technologies for straw burning as a form of renewable energy production (including in the form of whole bales) [19,20]. Straw has a higher specific calorific value than brown coal (4.9 kWh.kg^{-1} of dry matter or 4.0 kWh.kg^{-1} of straw with a moisture content of 15%) [21]. It is used as a heating element in many countries, not only for environmental reasons, but also because of its low cost [22,23]. The burning of straw is influenced by its chemical nature—a natural organic polymer. Xie et al. [24] carried out a comparative analysis of thermal oxidative decomposition and fire characteristics for different straw powders using thermogravimetry and cone calorimetry. The fire characteristics of straw are very similar to those of wood. The thermal degradation of straw starts at 270 °C, occurs in two steps, and the final temperature is 600 °C [24].

Straw breaks down (Figure 1) and in this case increases the area of the flammable substance that can enter the oxygen reaction, which is significant because the presence of an oxidizer is crucial to continued burning.

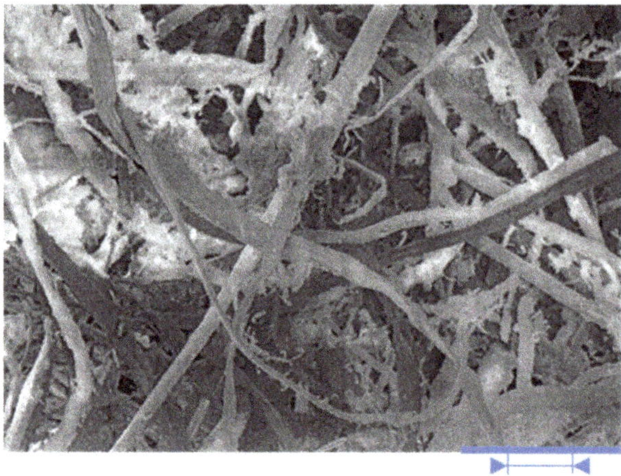

Figure 1. Light microscopy images of straw fibers with magnification 200×. Legend: Blue line presents a size of 100 microns (μm) in 2D layout.

Eco-panels are formed from pressed straw enclosed in a surface material; the pressed, tightly packed nature of the straw prevents the oxidative agent from reacting with the thermally degraded layer.

The results of several tests and measurements of straw structures intended to define its fire characteristics show that straw constructions with a proper coating have excellent values of fire resistance [16,25–29]. Surface treatments on straw walls (e.g., clay plaster, plasterboard cladding, and lime plaster) can increase the fire resistance of structural components [29,30]. Field and laboratory tests carried out by Theis [31] show plastered bale walls to be highly resistant to fire damage, flame spread, and combustion [31].

Cardboard treatment represents a decisive factor. Producers of eco-panels assert that there is a self-extinguishing effect [10,16] because of the highly pressed straw in the eco panel core, which contains a minimum amount of air. If an eco-panel burns, the paper burns first (Cardboard) followed by the straw, which slows down or even stops the process [32].

The improvement of the fire performance of straw and other similar materials can be achieved in two ways—by physical and chemical retardation. Physical retardation is based on regulating the size of the input material and the output density of the material. Chemical retardation is possible by applying a retardant to the input material, thus making a "new" product out of it, or by adding a retardant during the manufacturing process. All of the above-mentioned procedures can be carried out for eco panels.

The purpose of this article is to monitor the fire behavior and define the fire properties of eco-panels, defined as a pressed straw core with cardboard surface treatment used in wooden constructions, when exposed to flame. The article also deals with non-fire-retarded material in order to obtain input parameters for potential further observations.

2. Materials and Methods

2.1. Test Samples

Two groups of samples, 30 mm thick, were used in the experiment (Table 1). The first set of samples were 50 × 100 mm in size and included 3 groups of 5 samples, marked A, B, and C (Figure 2c). The second group, marked D, were samples 100 × 200 mm in size (Figure 4). All samples were conditioned for 24 h in a burning laboratory and their weight was determined. As the sample size increased, the weight differences became more significant (Table 1).

Table 1. Test samples.

Series	Group	Composition	Dimensions (mm)	Weight (g)	Burner Position	Test Time (min)
1.	A		50 × 100	101.74 ± 2.381	Lateral side of panel	5
	B	Straw Cardboard		103.74 ± 1.785	Edge of panel	5
	C			99.74 ± 1.998	Bottom of panel	5
2.	D		100 × 200	429.14 ± 34.180	Below 45°	10

The experiment was carried out in laboratory conditions with an ambient temperature of 21.5 °C, air flow 0.02 m·s^{-1}, and ambient humidity 57%.

2.2. The Ignitability Fire Test by a Small-Time Attack Flame of Eco Panel

The experiments were carried out in the fire-chemical laboratory of the Faculty of Security Engineering at the University of Zilina in Zilina. Samples A, B, and C were tested on a non-certified device (Figure 2a). The ignitability fire test by a small-time attack flame of construction products was carried out in the special fire box. The test sample was mounted in a vertical position.

Samples A, B, and C were exposed to a small ignition initiator, a flame, for 5 min. The distance between the flame and the surface of the sample was maintained at 15 mm during experiments. The position of the burner, however, changed for each set of samples (see Table 1 and Figure 2b).

The samples of group A were exposed to a flame perpendicular to the lateral side of the eco panel, with the flame touching the surface of the cardboard of the eco-panel carton, approximately 1 cm from the bottom edge (illustration Figure 3a,d).

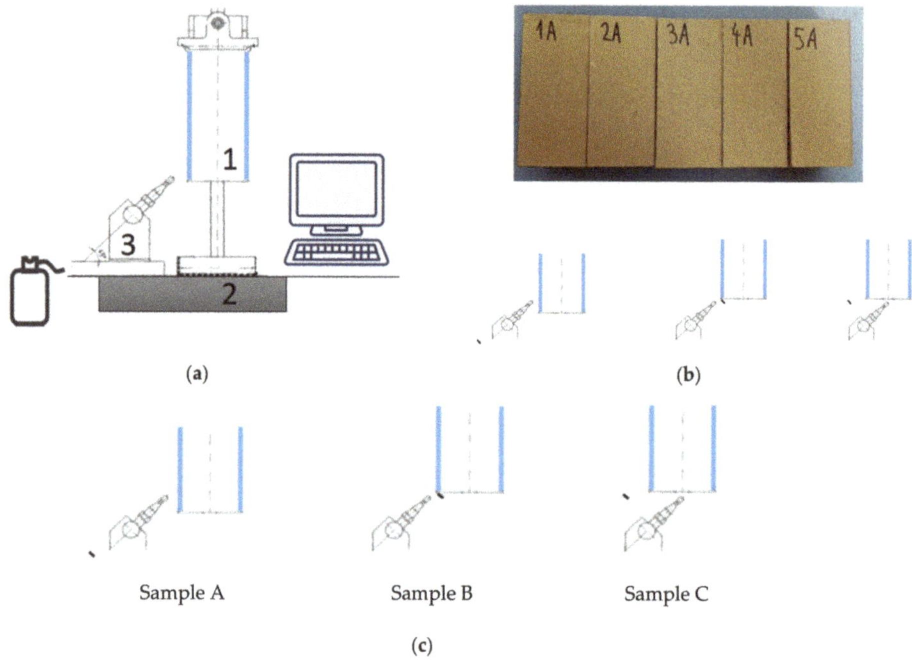

Figure 2. (**a**) Test equipment scheme; (**b**) First set of samples (A); (**c**) position of the burner relative to sample A, to sample B, to sample C. Legend for Figure 2. (**a**): 1—test sample, blue lines represent cardboard, 2—scales monitoring mass loss, 3—flame burner, in position of sample A.

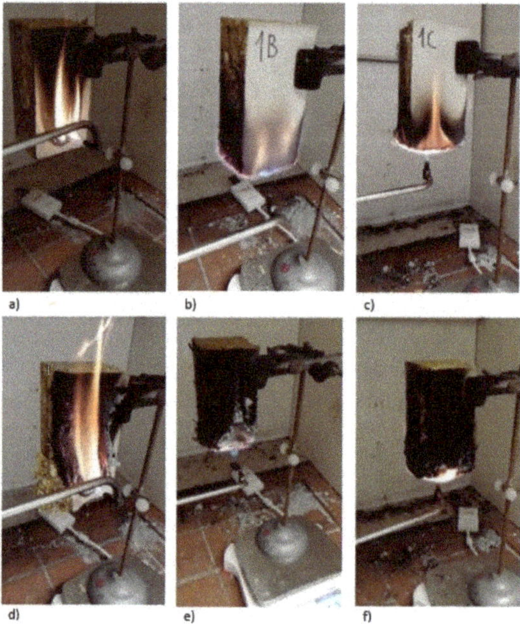

Figure 3. Location of the burner in the experiments in the first minute (**a**–**c**) and 3rd minute (**d**–**f**) for samples A (**a,d**); B (**b,e**); C (**c,f**).

Group B samples were exposed to a flame on the edge of the eco panel, such that the flame acted on both the cardboard and pressed straw (example Figure 3b,e).

Group C samples were exposed to flame from below, directly onto the pressed straw (demonstration Figure 3c,f).

2.3. Test Evaluation

The test evaluated flame spread along the surface of the sample to 100 mm (A, B, and C) or 200 mm (D) vertically from the point of contact with the flame. The results indicate whether the flame spread over the sample (Yes) or it did not (No). Furthermore, the time for which flame spread occurred was recorded, as was mass loss of the samples.

Two test methods based on different evaluation principles were used to assess the experiment. The first focused on the propagation of the flame over the surface of the material, so the samples were in a vertical position. In this case, the position of the source that ignited the material was important. The fire source was the flame of a gas burner (pure propane) with a flame length of 20 mm. As previously mentioned, in order to achieve a comprehensive evaluation of the material, testing of its ignition in several positions of the flame and the sample was carried out (see Table 1 and Figure 2). Position A represents the effect of the flame on the surface of the sample (40 mm) from its lower edge (Figure 3a,d). Position B represents the action of the flame on the edge of the material, so that the flame acted on both the paperboard and the pressed straw (example Figure 3b,e). In Position C, the flame acted on the center of the sample within its width and thickness. Group C samples were exposed to the flame from below directly on the pressed straw (demonstration, Figure 3c,f). These burner positions revealed the weaknesses in the material. Samples A, B, and C were exposed to a small ignition initiator, a flame, for 5 min. During the experiments, the distance between the flame and the sample surface was maintained at 15 mm.

The second experiment monitored the actual burning (flame propagation) using the D samples (200 × 100 × 10 mm). The higher intensity flame from the Bunsen burner acted on the center of the sample along its length and width, and the sample was placed at an angle of 45° to the horizontal plane (see Figure 4). The flame loading time of the sample was 10 min. The flame size was 100 mm and the position of the burner orifice from the center of the sample was 90 mm. Evaluation criteria of individual experiments are stated in Section 2.4.

(a)

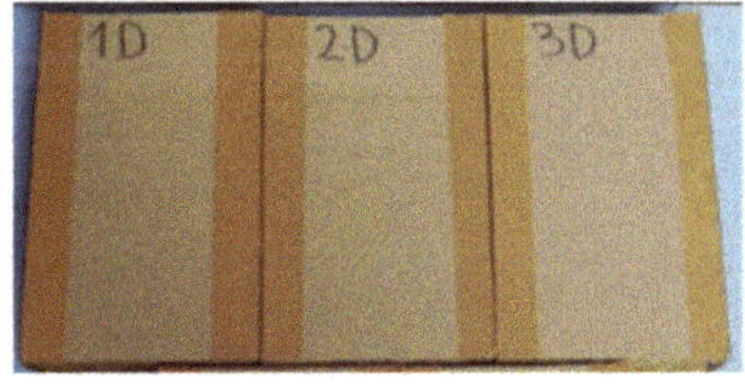

(b)

Figure 4. (a) Treatment of the experiment for group D samples. (b) Samples D.

2.4. Weight Loss and Relative Burning Rate

When the samples were exposed to heat, we observed and recorded weight loss in 10 s intervals. Relative weight loss was calculated according to the relation (1) [33]:

$$\delta_m(\tau) = \frac{\Delta m}{m(\tau)} \cdot 100 = \frac{m(\tau) - m(\tau + \Delta\tau)}{m(\tau)} \cdot 100 \, (\%) \tag{1}$$

where $\Delta_m(\tau)$ is relative weight loss in time (τ) (%), $m(\tau)$ is sample weight in time (τ) (g), $m(\tau + \Delta\tau)$ is sample weight in time ($\tau + \Delta\tau$) (g), Δm is weight difference (g).

Relative burning rate has been determined according to the following relations (2) [33] and (3) [33]:

$$v_r = \left| \frac{\partial \delta_m}{\delta \tau} \right| (\%/s) \tag{2}$$

or numerically

$$v_r = \frac{|\delta_m(\tau) - \delta_m(\tau + \Delta\tau)|}{\Delta\tau} (\%/s) \tag{3}$$

where v_r is relative burning rate (%/s), $\Delta_m(\tau)$ is relative weight loss in time (τ) (%), $\Delta_m(\tau + \Delta\tau)$ is relative weight loss in time ($\tau + \Delta\tau$) (%), $\Delta\tau$ is time interval where the weights are subtracted (s).

2.5. Mathematical and Statistical Processing and Evaluation of Results

To evaluate the influence of burner position on the thermal degradation of the samples using mass loss, the results obtained were subjected to statistical analysis. The results regarding mass loss were statistically evaluated by a single-factor variation analysis (ANOVA) using the LSD test (confidence intervals of 95%, 99%, using STATGRAPHICS version 18/19 software), where the factor of influence was the contact point flame vs. sample based on the different positions of the burner.

3. Results

The experimental data obtained using mass loss in relation to exposure time (Figure 5) were used to calculate the burning rate.

All variants exposed to flame ignited. The variability of the results depend on the straw sample and its preparation method (such as applying glue on cardboard, etc.). The flame position and the point of contact between the flame and the sample had an effect on the results. The mass loss courses differed (Figure 5a,c,e). This influenced the change in mass loss, most notably for the sample B (Figure 6), in which the flame was located on the edge of the panel, so both the straw and the cardboard were exposed to flame.

Samples of group C burned in a relatively similar way; direct exposure of the straw to flame caused the slowest burning progress and the lowest rates of fire propagation (Figure 5e). During the 5-min test, the samples burned halfway (Figure 5f) and retained their shape.

When the eco panel burns, a lower mass loss occurs compared to massive wood [34]. Maximum mass loss for sample B is 10.05% (Figure 6).

The burning rate is determined as the mass loss over time. The time-dependent course of the burning rate is confirmed by the most intense burning behavior of sample B. The lowest burning rate was observed for the A samples (Figure 7). The assumption that group C samples would the lowest mass values was not confirmed. At the same time, the burning rate for these straw eco panels was one-tenth lower than the burning rate of wood. The Kacikova and Makovicka [35] survey, which monitored the burning rate of juniper, spruce, fir, and larch exposed to a heat initiator, resulted in values in the range of 0.038–0.080 (mg·s^{-1}). At the same time, the curves show a similar course to those showing the heat release rate (HRR) over time [24,36]. The eco-panel samples' burning rate curve have a regular course, and the rate of the burning rate in the initial phase does not have a sharp increase. The effect is probably slowed down due to the smooth surface of the cardboard. Subsequently, the case burned through and the straw started to burn (150 s), and there was a drop in the burning rate due to a loss of homogeneity of the combustible material.

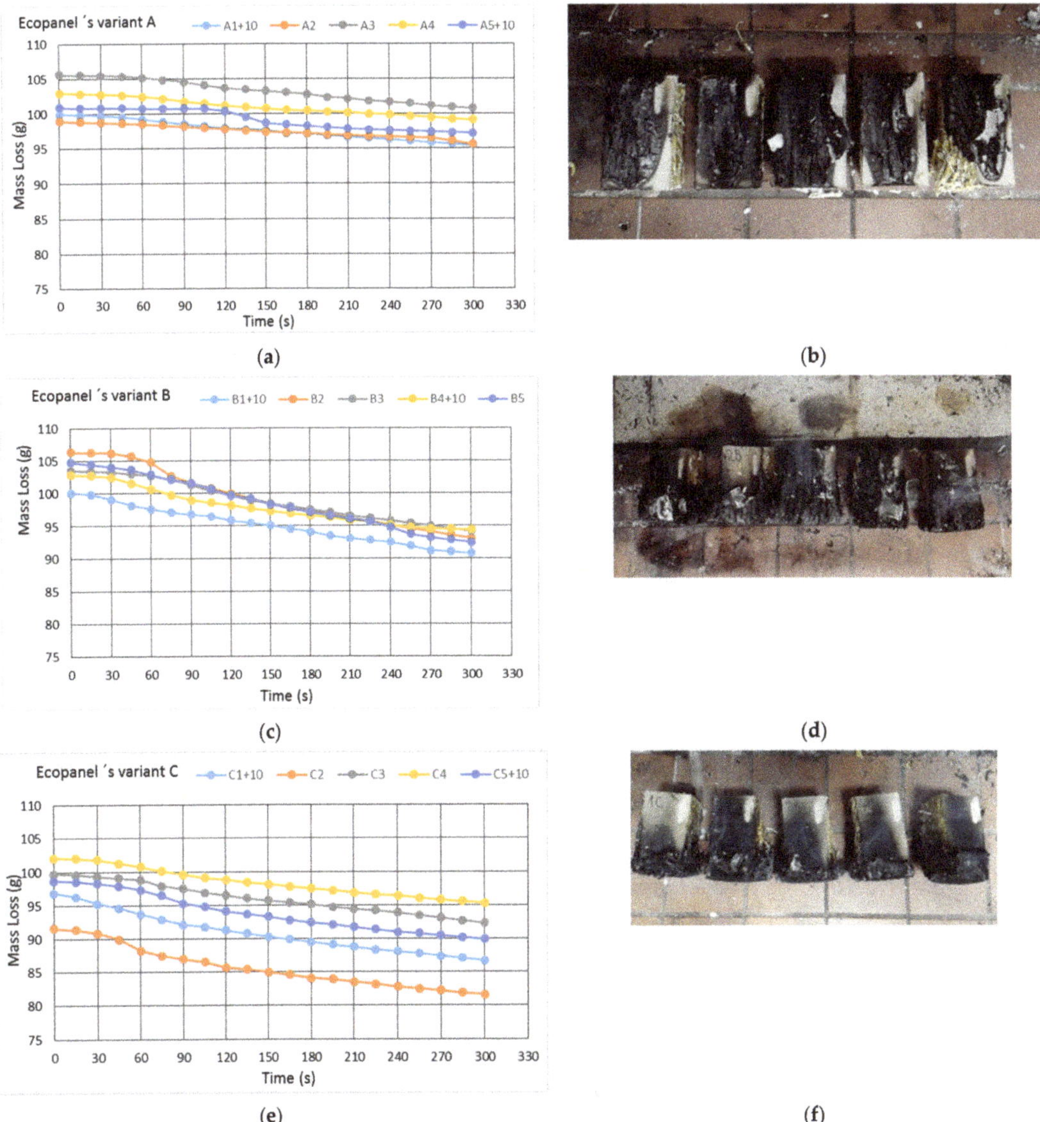

Figure 5. Graphical dependences of mass loss during experiment of (**a**) Group A samples, (**c**) Group B samples, (**e**) Group C samples. Samples after the experiment (**b**) Group A samples, (**d**) Group B sample, (**f**) Group D samples. Exposure time was 5 min.

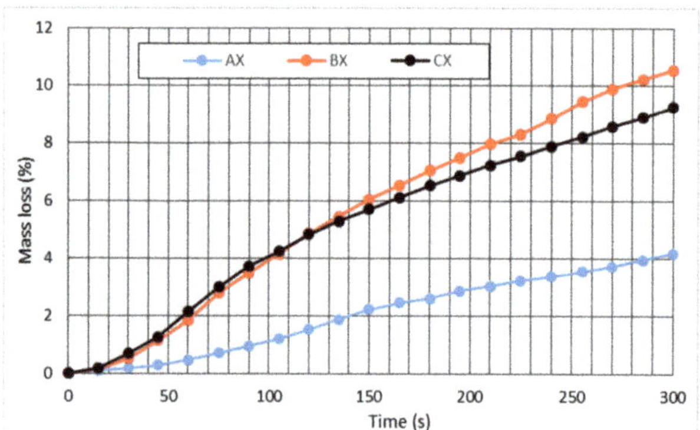

Figure 6. Mutual comparison of the increase in mass loss of eco panels in groups A, B, and C based on time of burning. Note: The value X represents the average value of all samples in each group.

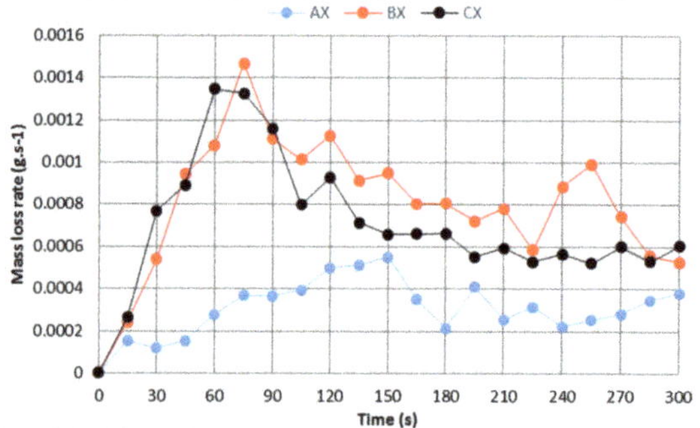

Figure 7. This is a figure. Schemes follow the same formatting. Note: The value X represents the average value of all samples in each group.

The application of a single-factor ANOVA (STATGRAPHICS version 18/19 software) analysis did not confirm a significant difference in ignition time based on panel thickness. (Table 2; Figure 8).

Table 2. One-way ANOVA for statistical evaluation of the influence of the burner's position on the burning rate (weight loss) of the sample when Eco-panels in groups A, B, and C were exposed to a flame.

Sample	Number	Mean	Standard Deviation	Variance	Standard Error
Time	100	157.7	86.9299	7559.8181	8.6929
A	100	2.1363	1.4263	2.0343	0.1426
B	100	0.0008	0.0004	1.8864	4.3432
C	100	5.1528	2.9357	8.6183	0.2935

Source	Df	Sum of Squares	Mean Square	F Value	*p* Value
Model (Between groups)	3	1,804,842.948	601,614.3160	318.0002	4.6151
Error	369	749,179.6158	1,891.8677		
Total	399	2,554,022.5639			

Null Hypothesis: Means of all selected data sets are equal. Alternative Hypothesis: Means of one or more selected datasets are different. At the 0.05 level, the population means are significantly different.

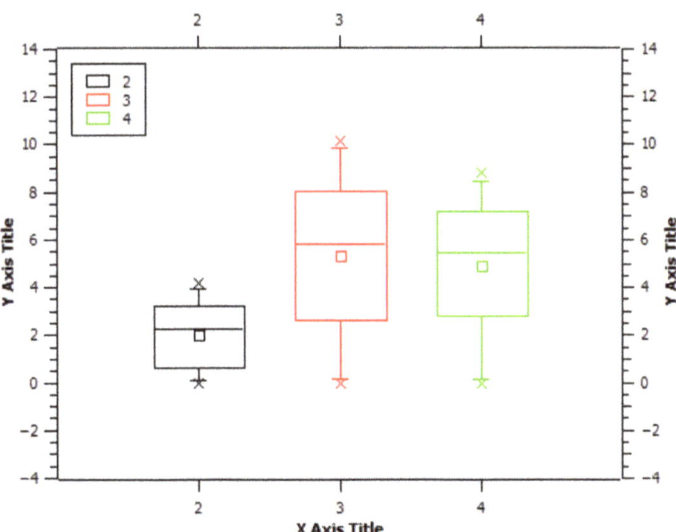

Figure 8. Graphical representation of the statistical evaluation of the influence of the burner's position on the mass loss of the sample when eco-panels in groups A, B, and C were exposed to the flame. Legend: X Axis Title = Samples. (2-Group A samples; 3-Group B samples; 4-Group C samples.) Y Axis Title = Mass Loss.

Statistical Results: The Null Hypothesis Applies

Another aim of the research was to follow the behavior of samples of larger dimensions and simulate the position of roofing sleepers in wooden houses. The samples were placed at an angle of 45° and the burner was placed into the center of the sample. The burning time was 10 min, the standard time of fire-fighting units' arrival. The results of the burning behavior of samples in group D show reduced mass loss and maintain the original sample dimensions (Figure 9). During the experiment, the flame operated on the surface and did not reach the inner part of the panel. In all samples, the cardboard layer burned through and straw began to burn. The process was slower in comparison to the A, B, and C samples, and the maximum value of weight loss was 6.4% (Figure 9c). At the same time, the burning rate decreased by one-tenth compared to A, B, and C samples.

Several authors [11,16,37] conducted research into eco-panels as construction elements according to the Fire Reaction Test methodology. They used the ignitability fire test by a small time attack flame and concluded that crushed straw might not be classified into class E according to its reaction to fire [16]. Class E is a category of products which are capable of resisting a small flame without substantial flame spread, for a short time [38,39]. The Class D category, however, is defined as products which satisfy the criteria for class E and are also capable of resisting a small flame without substantial flame spread for a longer period. In addition, they are also capable of undergoing thermal attack by a single burning item with sufficiently delayed and limited heat release [39]. Whether crushed straw could be classified into Class D must be confirmed by another series of fire tests.

Eco-panel, a straw building element, is characterized by a potential generation of dangerous gases. This is caused by imperfect burning of straw [40,41]. Jenkins [42] presents the research of pollutant emission factors for open field burning of wood and rice straw as comparable. They differ only in the production of CO; in wood it is 5.54%, of dry fuel, and in rice straw 3.22% of dry fuel.

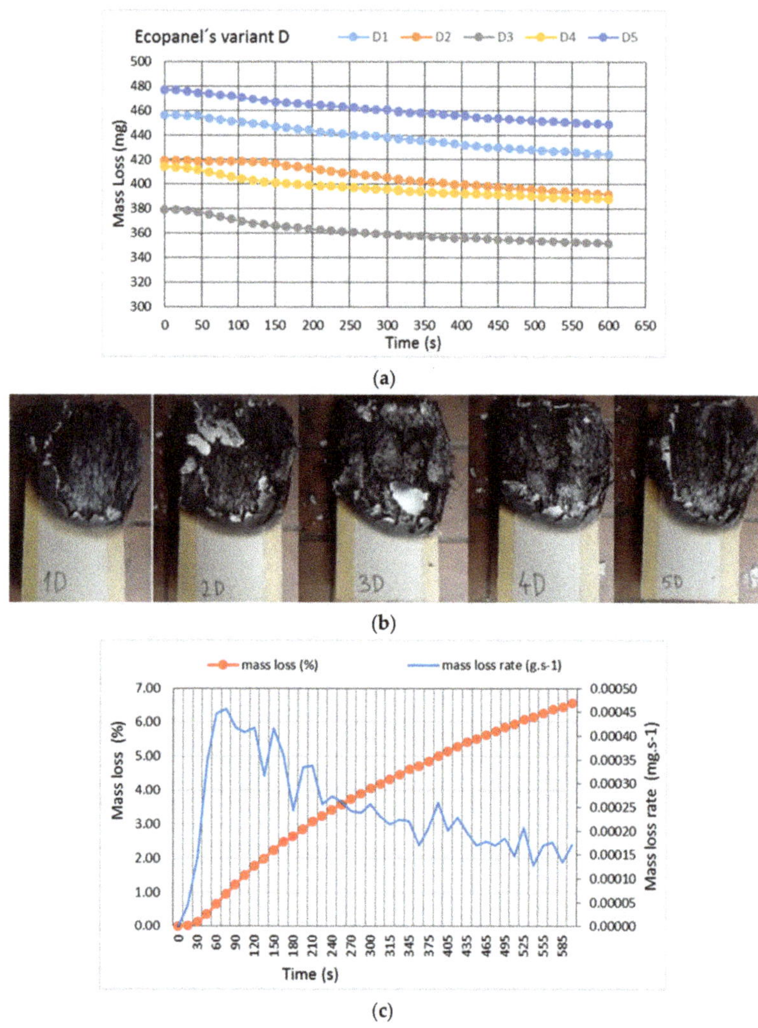

Figure 9. (**a**) Weight loss of D samples over time; (**b**) Group D samples after the experiment, exposure time 10 min; (**c**) The increase in mass loss and the burning rate of D samples. All charts display average values of all the samples.

4. Conclusions

The tests were performed on eco panel board samples with small dimensions and the results correspond to their dimensions. Eco panel (straw with cardboard casing) thermally degrades and burns with a luminous flame. As a result of exposure to the flame, initiation, and burning occur at rates slower than the standard values of the burning of wood (as a structure). A side effect is the instability of the eco panel. The samples gradually disintegrate, and parts of the straw fly off. One of the priority conclusions is the insufficient size of test specimens A, B, and C. For an objective assessment of the specimens, minimum dimensions of the tested sample 180 × 90 × 40 mm as specified in EN 13501-2 + A1 [43] are necessary. As sample D was twice as big as the other samples, it had a significantly different, slowed, burning (slowed down) course over a doubled time of exposure. In tests with large dimensions of eco panel boards (2.8 × 3 m) in wooden building structures, they showed a fire resistance of 120 min, which is declared sufficient by the Fire resistance test report [43].

The presentation of one part of the research results, which deals with monitoring the fire characteristics of natural materials applicable in building structures, shows the risk of thermal decomposition of eco panels, but above all, shows the importance of size, shape, and positioning of the eco panel in the structure in order to increase fire resistance.

Author Contributions: Conceptualization, L.M.O.; methodology, L.M.O.; software, I.M.; validation, L.M.O. and I.M.; formal analysis, I.M. and S.J.; investigation, L.M.O.; resources, I.M.; data curation, I.M.; writing—original draft preparation, L.M.O.; writing—review and editing, L.M.O.; project administration, J.B.; funding acquisition, I.M. All authors have read and agreed to the published version of the manuscript.

Funding: This research received no external funding.

Institutional Review Board Statement: Not applicable.

Informed Consent Statement: Not applicable.

Data Availability Statement: Not applicable.

Conflicts of Interest: The founding sponsors had no role in the design of the study, in the collection, analyses, or interpretation of data; in the writing of the manuscript and in the decision to publish the results.

References

1. Al Shawaf, Z. *An Experimental Study of Dry Onion Skins as Renewable Materials for Interior Finishes and Their Impact on Indoor Environment. Dissertation Work. Faculty of Engineering & IT*; The British University in Dubai: Dubai, United Arab Emirates, 2014; p. 103. Available online: https://bspace.buid.ac.ae/bitstream/1234/1000/1/2013133124.pdf (accessed on 29 September 2020).
2. Prečo Si Postaviť Dom Zo Slamy? Available online: https://www.stavebnictvoabyvanie.sk/stavebnictvo/4218-preco-si-postavit-dom-zo-slamy (accessed on 29 September 2020). (In Slovakian)
3. Secchi, S.; Asdrubali, F.; Cellai, G.; Nannipieri, E.; Rotili, A.; Vannucchi, I. Experimental and environmental analysis of new sound-absorbing and insulating elements in recycled cardboard. *J. Build. Eng.* **2016**, *5*, 1–12. [CrossRef]
4. Ekopanely-Stavební Desky Ze Slámy. Available online: https://www2.zf.jcu.cz/~{}moudry/databaze/Ekopanely.htm?fbclid=IwAR2OscJd9wM2Bhxm5tJBursy0YvOQtXTCxV2AvRQzXBCU6JeV07gVTWiDh0 (accessed on 18 April 2018). (In Czech)
5. Spottiswoode, A.J.; Bank, L.C.; Shapira, A. Investigation of paperboard tubes as formwork for concrete bridge decks. *Constr Build Mater.* **2012**, *30*, 767–775. [CrossRef]
6. Kadlicová, P.; Gašpercová, S.; Makovicka Osvaldova, L. Monitoring of Weight Loss of Fibreboard during Influence of Flame. *Procedia Eng.* **2017**, *192*, 393–398. [CrossRef]
7. Antov, P.; Savov, V.; Neykov, N. Possibilities for Manufacturing Insulation Boards with Participation of Recycled Lignocellulosic Fibres. *Manag. Sustain. Dev.* **2019**, *75*, 72–76. [CrossRef]
8. Delgado, B.; López González, D.; Godbout, S.; Lagacé, R.; Giroir-Fendler, A.; Avalos Ramirez, A. A study of torrefied cardboard characterization and applications: Composition, oxidation kinetics and methane adsorption. *Sci. Total Environ.* **2017**, *593–594*, 406–417. [CrossRef]
9. EKOPANEL-Slaměný Lisovaný Panel. Available online: https://www.prirodnistavba.cz/ekopanel-slameny-lisovany-panel-3426.html (accessed on 17 April 2020). (In Czech)
10. Cai, Z.; Robert, J.R. *Mechanical Properties of Wood-Based Composite Materials. Wood Handbook: Wood as an Engineering Material: Chapter 12*; General Technical Report FPL, GTR-190; U.S. Dept. of Agriculture: Madison, WI, USA, 2010; pp. 12.1–12.12.
11. Yang, H.; Kim, D.; Kim, H. Rice straw-wood particle composite for sound absorbing wooden construction materials. *Bioresour. Technol.* **2003**, *86*, 117–121. [CrossRef]
12. Sadzevicius, R.; Gurskis, V.; Ramukevičius, D. Sustainable construction of agro-industrial buildings from straw panels. In Proceedings of the Rural Development 2015, Kaunas, Lithuania, 23–24 April 2015.
13. Torun, G.; Korkut, Ö. Preparation of Cement Based Composites and Cellulosic Panels from Barley Straw for Thermal Insulation. *GU J. Sci.* **2017**, *30*, 31–32. Available online: https://dergipark.org.tr/tr/download/article-file/290214 (accessed on 29 September 2020).
14. Antov, P.; Savov, V.; Krišťák, Ľ.; Réh, R.; Mantanis, G.I. Eco-Friendly, High-Density Fiberboards Bonded with Urea-Formaldehyde and Ammonium Lignosulfonate. *Polymers* **2021**, *13*, 220. [CrossRef]
15. Antov, P.; Krišťák, L.; Réh, R.; Savov, V.; Papadopoulos, A.N. Eco-Friendly Fiberboard Panels from Recycled Fibers Bonded with Calcium Lignosulfonate. *Polymers* **2021**, *13*, 639. [CrossRef]
16. Teslík, J.; Vodičková, M.; Kutilová, K. The Assessment of Reaction to Fire of Crushed Straw. *Appl. Mech. Mater.* **2016**, *824*, 148–155. [CrossRef]

17. Štenclová, E. Spúšť v Žiline. Most Poškodili Horiace Pivné Prepravky, Fľaše A Slama (Trigger in Žilina. The bridge was Damaged by Burning Beer Crates, Bottles and Straw). Available online: https://spravy.pravda.sk/regiony/clanok/512806-poziar-budovy-pivnych-prepraviek-flias-a-slamy-poskodil-zilinsky-most (accessed on 21 May 2019).
18. Yang, Y.; Liu, J.; Jin, S.; Li, T. Experimental study about influence of particle size and oxygen atmosphere on straw powder's combustion characteristics. In Proceedings of the 2011 International Conference on Electrical and Control Engineering, Yichang, China, 16–18 September 2011; pp. 3470–3473. [CrossRef]
19. Pepich, Š. Slama Ako Zdroj Energie Z Poľnohospodárstva (Straw as a Source of Energy from Agriculture). Available online: http://old.agroporadenstvo.sk/oze/biomasa/slama.pdf (accessed on 17 April 2020). (In Slovakian)
20. Saidur, R.; Abdelaziz, E.; Demirbas, A.; Hossain, M.; Mekhilef, S. A review on biomass as a fuel for boilers. *Renew. Sustain. Energy Rev.* **2011**, *15*, 2262–2289. [CrossRef]
21. Pepich, Š. Pilotný Project Energetického Využitia Biomasy. Available online: http://www.tsup.sk/files/vyuzitie_poln.biomasy_na_energet.ucely.pdf (accessed on 1 December 2010). (In Slovakian)
22. Hajinajaf, N.; Mehrabadi, A.; Tavakoli, O. Practical strategies to improve harvestable biomass energy yield in microalgal culture: A review. *Biomass Bioenergy* **2021**, *145*, 105941. [CrossRef]
23. Schnorf, V.; Trutnevyte, E.; Bowman, G.; Burg, V. Biomass transport for energy: Cost, energy and CO_2 performance of forest wood and manure transport chains in Switzerland. *J. Clean. Prod.* **2021**, *293*, 125971. [CrossRef]
24. Xie, T.; Wei, R.; Wang, Z.; Wang, J. Comparative analysis of thermal oxidative decomposition and fire characteristics for different straw powders via thermogravimetry and cone calorimetry. *Process. Saf. Environ. Prot.* **2020**, *134*, 121–130. [CrossRef]
25. Růžička, J. Požární Odolnost Obvodových Stěn Pro Pasivní Domy S Využitím Slaměných Balíků Jako Tepelné Izolace. Available online: https://stavba.tzb-info.cz/obalove-konstrukce-nizkoenergetickych-staveb/8974-pozarni-odolnost-obvodovych-sten-pro-pasivni-domy-s-vyuzitim-slamenych-baliku-jako-tepelne-izolace (accessed on 27 August 2012). (In Czech)
26. Růžička, J.; Pokorný, M. Požární Odolnost Obvodových Stěn NED, PD z Přírodních a Recyklovaných Stavebních Materiálů. *Stavebnictví* **2011**, *11*, 34–39. (In Czech)
27. Kozłowski, R.; Władyka-Przybylak, M. Flammability and fire resistance of composites reinforced by natural fibers. *Polym. Adv. Technol.* **2008**, *19*, 446–453. [CrossRef]
28. Chybík, J. *Přírodní Stavební Materiály (Natural Building Materials)*; Grada Publishing: Prague, Czech Republic, 2009; p. 272. (In Czech)
29. Pokorný, J.; Kučera, P.; Vlček, V. Specific knowledge in assessment of local fire for design of building structures. *Adv. Mat. Res.* **2014**, *1001*, 362–367. [CrossRef]
30. Hýsková, P.; Hýsek, Š.; Schönfelder, O.; Šedivka, P.; Lexa, M.; Jarský, V. Utilization of agricultural rests: Straw-based composite panels made from enzymatic modified wheat and rapeseed straw. *Ind. Crops Prod.* **2020**, *144*, 112067. [CrossRef]
31. Theis, B. Straw Bale Fire Safety. Available online: http://www.naturalbuildingcoalition.ca/Resources/Documents/Technical/strawbale_fire_safety.pdf (accessed on 30 July 2003).
32. Sobotka, M. The Choice of a Suitable Material for the Building of the Detached House; ČVUT Praha. Available online: http://stretech.fs.cvut.cz/2014/sbornik_2014/zdar%20nad%20sazavou_sobotka-volba%20materialu.pdf (accessed on 30 June 2014). (In Czech)
33. Makovicka Osvaldova, L. Experiment Description. In *Wooden Façades and Fire Safety*; SpringerBriefs in Fire; Springer: Cham, Switzerland, 2020. [CrossRef]
34. Weisberger, J.M.; Richter, J.P.; Mollendorf, J.C.; DesJardin, P.E. An emissions-based fuel mass loss measurement for wood-fired hydronic heaters. *Biomass Bioenergy* **2020**, *142*, 105731. [CrossRef]
35. Kačíková, D.; Makovická, L.M. Wood burning rate of various tree parties. *Acta Fac. Xylologiae* **2009**, *51*, 27–32. Available online: https://df.tuzvo.sk/sites/default/files/04-1-09-kacikova-makovicka-osvaldova.pdf (accessed on 15 April 2021). (In Slovakian)
36. Hao, H.; Chow, C.L.; Lau, D. Effect of heat flux on combustion of different wood species. *Fuel* **2020**, *278*, 118325. [CrossRef]
37. Ferrandez-Garcia, C.C.; Garcia-Ortuño, T.; Ferrandez-Garcia, M.T.; Ferrandez-Villena, M.; Ferrandez-Garcia, C.E. Fire-resistance, physical, and mechanical characterization of binderless rice straw particleboards. *BioResources* **2017**, *12*, 8539–8549. [CrossRef]
38. Fire Classification. Available online: http://www.paroc.com/knowhow/fire/fire-classification (accessed on 12 June 2015).
39. European Committe for Standartion. *EN 13501-2+A1 2010. Fire Classification of Construction Products and Building Elements—Part 2: Classification Using Test Data from Resistance Fire Tests, Excluding Ventilation Services*; European Committe for Standartion: Brussels, Belgium, 2010.
40. Cao, G.L.; Zhang, X.; Wang, Y.; Zheng, F.C. Estimation of emissions from field burning of crop straw in China. *Chin. Sci. Bull.* **2008**, *53*, 784–790. [CrossRef]
41. Gadde, B.; Bonnet, S.; Menke, C.; Garivait, S. Air pollutant emissions from rice straw open field burning in India, Thailand and the Philippines. *Environ. Pollut.* **2009**, *157*, 1554–1558. [CrossRef]
42. Jenkins, B.B.; Miles, L.T.; Baxter, L.L. Combustion properties of biomass. *Fuel Process. Technol.* **1998**, *54*, 17–46. [CrossRef]
43. *Fire Resistance Test Report No Pr-18-2.204/En Loadbearing External Wall Made of Boards Ecopanel Eco 2 Boards*; Testing Laboratory No. 1206; Testing Laboratory: Veselí nad Lužnicí, Czech Republic, 2018.

Experimental Study of Oriented Strand Board Ignition by Radiant Heat Fluxes

Ivana Tureková [1], Iveta Marková [2,*] , Martina Ivanovičová [1] and Jozef Harangózo [1]

[1] Department of Technology and Information Technologies, Faculty of Education, Constantine the Philosopher University in Nitra, Tr. A. Hlinku 1, 949 74 Nitra, Slovakia; ivana.turekova@ukf.sk (I.T.); mabumb@gmail.com (M.I.); jozef.harangozo@ukf.sk (J.H.)

[2] Department of Fire Engineering, Faculty of Security Engineering, University of Žilina, Univerzitná 1, 010 26 Žilina, Slovakia

* Correspondence: iveta.markova@fbi.uniza.sk; Tel.: +421-041-513-6799

Abstract: Wood and composite panel materials represent a substantial part of the fuel in many building fires. The ability of materials to ignite when heated at elevated temperatures depends on many factors, such as the thermal properties of materials, the ignition temperature, critical heat flux and the environment. Oriented strand board (OSB) without any surface treatment in thicknesses of 12, 15 and 18 mm were used as experimental samples. The samples were gradually exposed to a heat flux of 43 to 50 $kW.m^{-2}$, with an increase of 1 $kW.m^{-2}$. At heat fluxes of 49 $kW.m^{-2}$ and 50 $kW.m^{-2}$, the ignition times are similar in all OSB thicknesses, in contrast to the ignition times at lower heat fluxes. The influence of the selected factors (thickness and distance from the heat source) was analysed based on the experimentally obtained data of ignition time and weight loss. The experimentally determined value of the heat flux density was 43 $kW.m^{-2}$, which represented the critical heat flux. The results show a statistically significant effect of OSB thickness on ignition time.

Keywords: OSB; heat flux density; ignition time; weight loss

Citation: Tureková, I.; Marková, I.; Ivanovičová, M.; Harangózo, J. Experimental Study of Oriented Strand Board Ignition by Radiant Heat Fluxes. *Polymers* **2021**, *13*, 709. https://doi.org/10.3390/polym1305 0709

Academic Editor: Antonios N. Papadopoulos

Received: 1 February 2021
Accepted: 21 February 2021
Published: 26 February 2021

Publisher's Note: MDPI stays neutral with regard to jurisdictional claims in published maps and institutional affiliations.

1. Introduction

Composite panel materials are important wood products [1]. Their production encompasses the utilization of wood of lower quality classes to obtain suitable materials with improved physical and mechanical properties [2]. Oriented strand board (OSB) belong to this group of products but essentially, these products are input materials in the furniture and construction industries [3].

The production of wood-based sheet materials utilizes wood of lower quality classes and chemically safe recyclates and generates materials with improved physical and mechanical properties compared to raw wood [4]. Oriented strand boards (OSBs) are defined in [5] as multilayer boards made of wood strands of a specific shape, thickness and adhesiveness. Strands in the outer layers are oriented parallel to the length or width of the boards. Strands in the middle layer or layers may be oriented randomly or generally perpendicular to the strands of the outer layers [5].

Oriented strand board, also known as OSB, waferboard, Sterling board or Exterior board and SmartPly, is a widely used engineered wood product formed by strands (flakes) of wood, often layered in specific orientations [6]. In appearance, it may have a rough and variegated surface with the individual strands (typically around 2.5 by 15 cm each) lying unevenly across each other. OSBs are cheap and strong boards, and this makes them excellent building material [7].

OSB is produced from thinner wood strands, which can be arranged better with respect to the direction of the wood fibres when adding the layers in longitudinal and transverse directions in the production flow. Strands have the longest length in the direction of the fibres [8]. Thinner debarked forest wood assortments, predominantly soft deciduous woods but also coniferous woods, are used as a raw material for OSB production. Great emphasis is placed on the preparation of strands. They are mainly produced by disc and drum chippers [9]. Thinner and longer strands are used for the surface layer, thicker and slightly shorter strands are used for the middle layers. Fine particles and dust are carefully sorted out. The strands are pressed under high pressure and temperature, using formaldehyde-based synthetic resins [10].

Most commonly used adhesives are conventional synthetic adhesives, both liquid and powder, including isocyanate-based adhesives [11]. OSB is categorized based on the purpose of its application. General-purpose boards (manufactured with a thickness of 9, 11, 15 or 18 mm) and boards for indoor furnishing (including furniture) in dry conditions are classified as OSB/1; load-bearing boards for use in dry conditions are classified as OSB/2; load-bearing boards for use in humid conditions are OSB/3; and heavy-duty load-bearing boards for use in humid conditions are OSB/4 [12].

OSB panel product in terms of physical and health hazards is unclassified according to safety data sheets [13]. OSB is the flammable composite material. OSB can be in contact with heat sources, and it will react to the effect of heat and temperature rise in its structure [14,15]. Wood materials are thermal insulators and do not conduct heat; hence there is a gradual process of thermal degradation, which can result in ignition and fire [16,17]. In the case of thermal stress, the strength of OSB decreases with increasing temperature and time of its action, while at higher temperatures the rate of this change is higher [18].

Flammability is defined as the ability of a sample to ignite under the action of an external thermal initiator and under defined test conditions according to [19]. According to International Organization for Standardization (ISO) 3261 [20], flammability is the ability of a material to ignite. Flammability is characterized by the ignition time of substances and materials, which depends on the ignition temperature, thermal properties of materials, sample conditions (size, humidity, orientation) and critical heat flux [21]. The definition of the term "ignition temperature" can be interpreted as the minimum temperature to which the air must be heated so that the sample placed in the heated air environment ignites, or the surface temperature of the sample just before the ignition point [22,23]. This interpretation was used as the basis of our research. The results were realized on tested equipment without a small burner flame, having only a radiated heat loading.

The aim of this article is to monitor the significant effect of heat flux density (from 43 to 50 $kW.m^{-2}$) and thickness (12 mm, 15 mm, 18 mm) of OSB on the ignition time and change in the weight of the sample. The change in heat flux was previously monitored and recorded for the purpose of validating equipment for the research of heat release loading.

At the same time, the critical temperature of the ignition point of the OSB was experimentally determined depending on the time of action of the radiant heat source and the intensity of the heat flux.

2. Materials and Methods

2.1. Experimental Samples

Samples of oriented strand board (OSB) without surface treatment, produced by Kronospan Jihlava (KRONOSPAN CR, spol. s r.o., Jihlava, Czech Republic) under the title OSB/3 SUPERFINISH ECO (Figure 1), were selected for the experiments. The OSBs used were multilayer boards made of flat strands of a specific shape and thickness. The strands in the outer layers were oriented parallel to the length or width of the board, and the strands in the middle layers were oriented randomly or were generally perpendicular to the lamellas of the outer layers. They were bonded with melamine formaldehyde resin and polymeric diphenylmethane diisocyanate (PMDI) (Shandong Shanshi Chemical Co Ltd, Zhangdian District, Zibo City, Shandong Province, China) and they were flat-pressed. The boards contained mainly a mixture of coniferous wood. Table 1 shows the physical and chemical properties and fire technical characteristics of the OSB with thickness of 10 mm to 18 mm [10,13].

(a) Example of OSB

(b) Example of experimental sample of OSB

Figure 1. Oriented strand board in (**a**) full piece and (**b**) prepared for measurement according to [24].

Samples of OSB were cut to specific dimensions (165 × 165 mm) according to ISO 5657: 1997 [24]. The density of the strand boards was determined according to Standard EN 323: 1996 [25].

Table 1. Physical and chemical properties and fire technical characteristics of the OSB with thickness of 10–18 mm [13].

Parameters	Notes	Values
Density (kg.m^{-3})		630 ± 10%
Humidity (%)		5 ± 12%
Bending Strength (N.mm^{-2})	Main Axis	20
	Secondary Axis	10
Modulus of Elasticity (N.mm^{-2})	Main Axis	3500
	Secondary Axis	1400
Swelling (%)		15
Thermal Conductivity (W.m^{-2}.K^{-1})		0.13
Formaldehyde Content (mg.100 g^{-1})		8
Flame Spread Index [1]		83.3
Reaction to Fire	Thickness 9 mm [2]	D-s2, d2
	Thickness 18 mm [3]	D-s1, d0
Class of Fire Reaction [26]		E—Eoderately Flammable

[1] Flame spread index was calculated from peak heat release rate (HRR), total heat release and time to sustained ignition by ASTM E 84 ([27] and [28]; [2] [29]; [3] [10].

The selected board materials were stored at a specific temperature (23 °C ± 2 °C) at relative humidity (50 ± 5%). Rantuch et al. [30] state that OSB moisture does not have a significant effect on the critical conditions under which they can be ignited, but significantly affects the time at which it occurs.

2.2. Experimental Procedures

The methodology can be divided into three parts:

1. Verification of test equipment; a small burner flame was not used as the secondary ignition source.
2. Determination of ignition time and weight loss depending on the selected level of heat flux density and thickness of board materials and on the distance of selected board materials from the ignition source.
3. Determination of the critical temperature during ignition of an OSB with a thickness of 15 mm.

2.2.1. Verification of Test Equipment

The beginning of the experimental measurement entailed verification of the test equipment according to ISO 5657: 1997 [24]. The aim of this verification was to monitor the temperature parameters of the cone calorimeter. The cone calorimeter must provide a heat flux in the range of 10 to 70 kW.m^{-2} in the centre of the hole in top plate and in a base plate which matches with the bottom side of the top plate (Figure 1b).

Verification of the actual condition of the cone calorimeter was performed using a SBG01-100 radiometer with a cooling sensor (Hukseflux, Delft, The Netherlands) and a METEX M-3890D digital multimeter (Manufacturer: Zebronics, India) with a USBVIEV program (SOFTONIC INTERNATIONAL S.A., Barcelona, Spain). Several experimental measurements, which included collecting voltage values from the multimeter and evaluating the results using the calibration curve specified in the standard, confirmed that at the selected temperature, the cone calorimeter indicated the corresponding heat flux densities in accordance with the previous measurements performed during test equipment calibration.

Based on the obtained values and compared results, a graphical dependence of the heat flux on the temperature of the cone calorimeter was constructed (Figure 2). The measurements were repeated with each thickness of the board material 5 times.

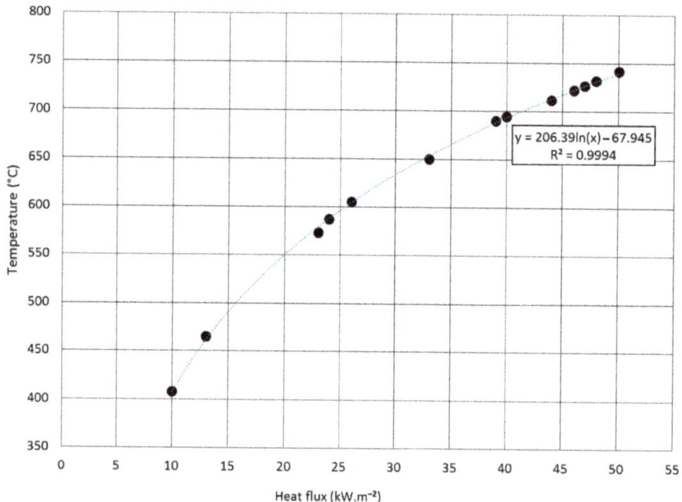

Figure 2. Heat flux dependence on cone calorimeter temperature.

2.2.2. Methodology for Determining Ignition Time and Weight Loss

The ignition time and weight loss depending on the selected level of heat flux density and thickness of the board materials, as well as on the distance of the selected board materials from the ignition source, were determined according to a modified procedure based on ISO 5657: 1997 [24]. This modification included a change of igniter. Ignition was caused only by heat flux, without the use of a direct flame (Figure 3a).

The samples were placed horizontally and exposed to a heat flux of 43 to 50 kW.m^{-2} by an electrically heated conical radiator (Beijing Global Trade Software Technology Co., Ltd., Chaoyang District, Beijing, China). Orientation experiments determined the minimum heat flux required to maintain flame combustion. The time-to-ignition value was recorded, while considering only the permanent ignition of the surface of the analysed sample when exposed to a selected level of heat flux density.

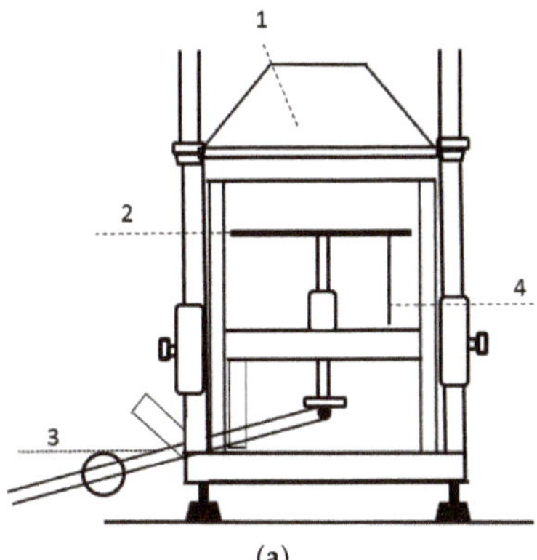

Figure 3. (a) Scheme of the equipment for determination of flammability of materials at a heat flux of radiant heat of 10–50 kW.m^{-2} according to [24]; (b) a photo of the real test equipment [31]. Legend: 1, heating cone; 2, board for sample; 3, movable arm; 4, connection point for recording experimental data.

The observed factors influencing the ignition time and weight loss were

- the thickness of the board material; and
- the density of the radiant heat flux.

2.2.3. Determination of the Critical Ignition Temperature

Determination of the critical ignition temperature was performed on a sample of OSB with a thickness of 15 mm. Test equipment according to [24] was supplemented by two thermocouples. These thermocouples were placed in the middle of the top and bottom surfaces of the board.

Test samples were exposed to four selected levels of heat flux density (44, 46, 48 and 50 kW.m^{-2}) for 300 s, which was repeated 5 times.

2.2.4. Statistical Processing of Data and Evaluation of Results

To evaluate the influence of the above-mentioned factors on the ignition temperature and weight loss, the obtained results were subjected to statistical analysis. The obtained results of the ignition and weight loss temperatures were statistically evaluated by one-way analysis of variance (ANOVA) using the Least Significant Difference LSD test (95%, 99% detectability level) (STATGRAPHICS software version 18/19, The Plains, VA, United States), with the use of board material thickness (12, 15 and 18 mm) and radiant heat flux density (from 43 to 50 kW.m^{-2}) as influence factors.

3. Results and Discussion

The course of the experiment (Figure 4) according to [24] confirmed the verified behaviour of the material in terms of the classification "reaction to fire D-s1, d0 by [10]". D-s1, d0 means "combustible materials – medium contribution to fire, with speed of emission absent or weak during combustion" by [26]. The priority of the experiment was to monitor the critical parameters of the ignition based on the change in board thickness (Table 2).

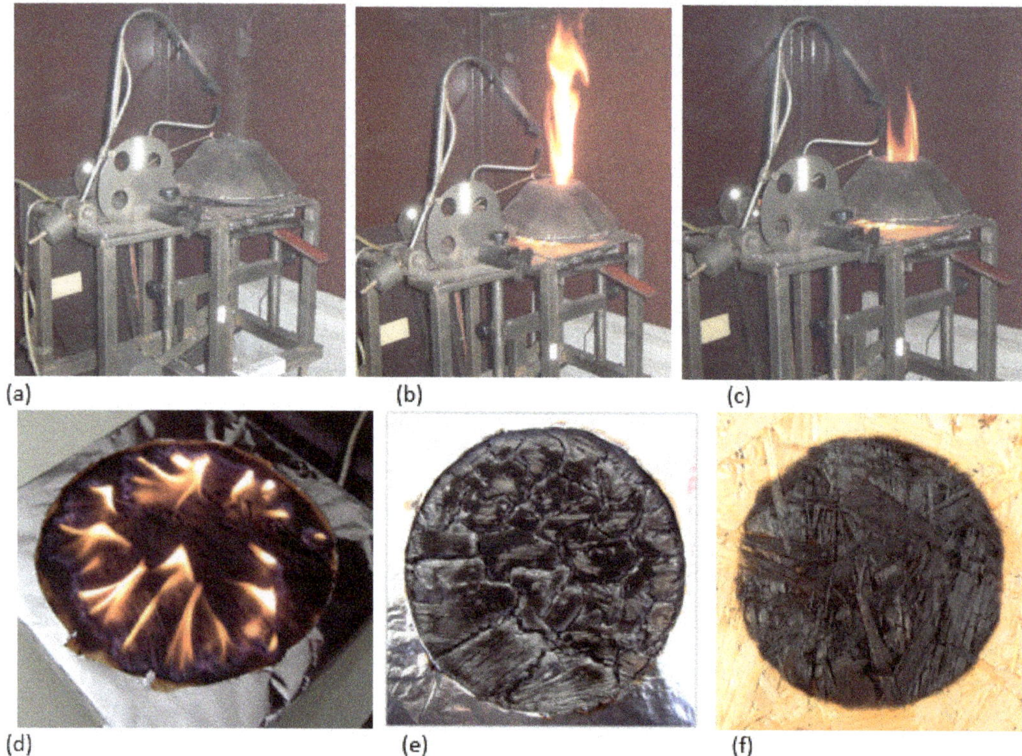

Figure 4. Demonstration of the course of burning of oriented strand boards (OSBs) after their ignition by radiant heat (**a**) ignition (100 s); (**b**) burning time of 160 s; and (**c**) 300 s. (**d**) A demonstration of the combustion process by 44 kW.m^{-2}. (**e**) A sample selected from the measuring device; specific distance of 12 mm, time of ignition of 57 s and density heat flux of 50 kW.m^{-2}. (**f**) Cooled sample 10 min after the experiment; sample thickness of 15 mm and ignition time of 108 s, 44 kW.m^{-2}.

3.1. Determination of Ignition Temperature and Weight Loss

For the purposes of our research, the initial value of the radiant heat flux, to which the selected OSBs were exposed, was experimentally set to 43 kW.m^{-2}. This value represented the critical heat flux for the selected samples. Critical heat flux is the heat flux between the minimum incident heat fluxes at which ignition occurs and maximum incident heat flux at which ignition does not occur. It can be used to evaluate the ignition ability. The critical heat flux is obtained experimentally by gradually exposing the samples to decreasing heat flux until ignition ceases [30]. The maximum value of the radiant heat flux, to which selected board materials were exposed, was 50 kW.m^{-2}. The heat flux gradually increased by 1 kW.m^{-2} (Table 2).

The effect of external heat flux on a sample is the incident energy upon its surface. Part of the energy is reflected (depending on the emissivity of the surface), part is transferred by conduction to deeper layers of the material (depending on its thermal conductivity) and the rest is absorbed by a thin layer on the surface, i.e., heating takes place. As the temperature of the sample increases, pyrolysis and thermal oxidation occur, as a result of which gaseous products are released into the environment [30]. Therefore, the time to ignition of the sample decreases with increasing external heat flux [31], which is confirmed by our results.

Table 2. Ignition time and weight loss in samples with different thicknesses using heat fluxes of 40 to 50 kW.m^{-2} at a distance of 20 mm.

Density of Radiant Heat Flux (kW.m^{-2})	Corresponding Temperature (°C) [1]	Thickness (mm)	Ignition Time (s)	Weight Loss (%)
43	700	12	107.4 ± 32.927	19.018 ± 0.742
		15	172.8 ± 68.271	16.528 ± 1.103
		18	170.0 ± 19.279	12.436 ± 0.402
44	710	12	80.80 ± 14.372	20.188 ± 1.210
		15	108.0 ± 31.093	16.092 ± 0.885
		18	140.0 ± 31.698	13.256 ± 0.745
45	720	12	86.4 ± 10.442	20.87 ± 0.889
		15	100.2 ± 21.673	17.026 ± 0.541
		18	111.2 ± 24.235	13.716 ± 0.303
46	724	12	84.4 ± 9.002	21.868 ± 0.879
		15	93.4 ± 21.767	17.272 ± 0.647
		18	98.8 ± 12.592	13.504 ± 0.228
47	727	12	67.08 ± 5.403	22.026 ± 0.908
		15	71.0 ± 8.671	17.5 ± 0.455
		18	103.6 ± 18.391	13.818 ± 0.266
48	730	12	58.60 ± 5.953	23.206 ± 0.505
		15	63.40 ± 7.116	18.366 ± 0.910
		18	77.60 ± 25.881	14.222 ± 0.826
49	735	12	62.20 ± 3.2497	23.578 ± 0.858
		15	63.20 ± 3.187	18.764 ± 0.571
		18	65.0 ± 11.436	14.678 ± 0.899
50	742	12	56.80 ± 2.039	24.302 ± 0.814
		15	59.40 ± 5.607	19.402 ± 0.586
		18	60.20 ± 5.741	14.846 ± 1.033

[1] Based on the graphical dependence in Figure 3.

An interesting result is the reduction of time differences in individual thicknesses at higher heat fluxes. Heat fluxes of 49 and 50 kW.m^{-2} offered almost equal times in all thicknesses. Differences in ignition times decreased with increasing heat flux, and the ignition differences between the individual thicknesses decreased. The maximum difference of ignition time between thicknesses (67.4 s) was at 43 kW.m^{-2} and the minimum (3.4 s) was at 50 kW.m^{-2}.

It can therefore be concluded that with increasing heat flux value, the thickness of the OSB has no significant effect on the ignition time. The increase in weight loss was with increasing heat flux, namely in 12 mm by 5.3%, in 15 mm by 2.9% and in 18 mm by 2.41%.

With increasing heat flux, there was an increase in weight loss of 12 mm by 5.3%, 15 mm by 2.9% and 18 mm by 2.41%. It is possible to make the following assumption: smaller thicknesses of OSB due to the increase of heat flow have more intensive thermodegradation processes, which manifest in higher weight losses.

Mitterová and Garaj [32] studied the weight loss and ignition time of an OSB caused by the action of radiant heat. Test samples were exposed to an infrared heater with a power of 1000 W for 600 s, while the distance of the samples from the surface of the radiating body was 30 mm.

The weight loss of the OSB reached 58.92%, and the ignition time was 53.4 s. In comparison with the values obtained by our research, the results are significantly different, because the distance between the emitters and the sample was relatively high, which affected the ignition time. The surprising result is more than 50% weight loss under these conditions.

The use of one-way ANOVA confirms the significant dependence of the ignition time on the board thickness (Table 3, Figure 5). The ANOVA table decomposes the variance of Col_4: Time- to-ignition into two components: a between-group component and a within-group component. The F-ratio, which in this case equals 4.31448, is a ratio of the between-group estimate to the within-group estimate. Since the P-value of the F-test is less than 0.05, there is a statistically significant difference between the mean Col_4 from one level of Col_2 to another at the 5% significance level. To determine which means are significantly different from which others, select Multiple Range (Table 3).

Table 3. ANOVA table for Col_4 Time-to-ignition by Col_2 thickness.

Source	Sum of Squares	Df	Mean Square	F-Ratio	P-Value
Between Groups	12,608.1	2	6304.03	4.31	0.0156
Within Groups	170,953	117	1461.13		
Total (Corr.)	183,561	119			

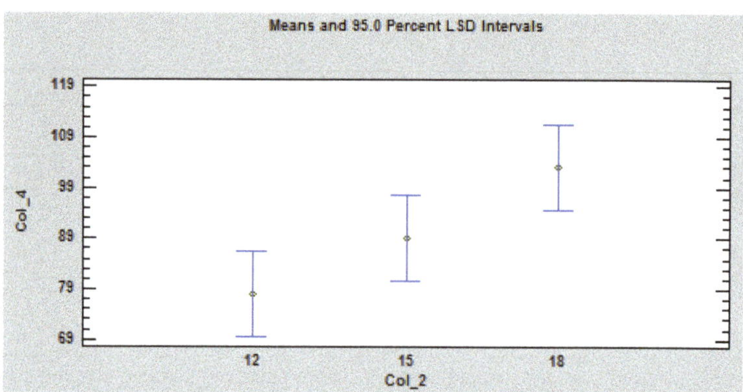

Figure 5. Graphical representation of the statistical evaluation of the influence of the board thickness on the ignition time under the action of the radiant heat source on the OSB. Legend: Col_2, thickness; Col_4, Time-to-ignition.

Concurrently, a significant dependence of the ignition time on the heat flux was confirmed (Table 4, Figure 6). The ANOVA table decomposes the variance of Col_4 into two components: Time-to-ignition and heat flux. The F-ratio, which in this case equals 17.5456, is a ratio of the between-group estimate to the within-group estimate. Since the P-value of the F-test is less than 0.05, there is a statistically significant difference between the means Col_4 Time-to-ignition from one level of Col_1 heat flux to another at the 5% significance level. To determine which means are significantly different from which others, select Multiple Range (Table 4).

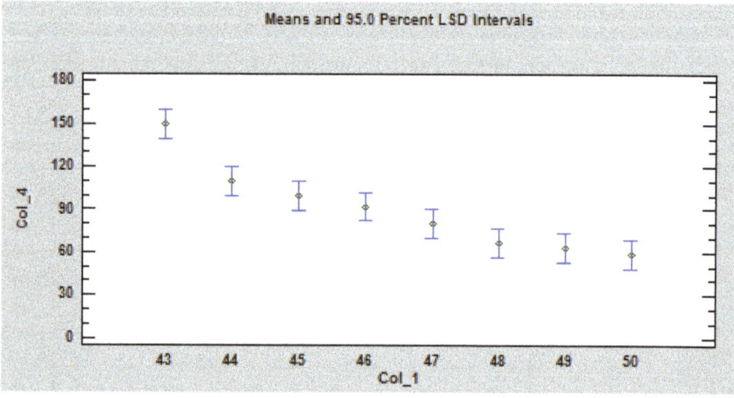

Figure 6. Graphical representation of the statistical evaluation of the influence of heat flux on ignition time under the action of a radiant heat source on the OSB. Legend: Col_1, heat flux; Col_4, Time-to-ignition.

Table 4. ANOVA table for Col_4 Time-to-ignition by Col_1 heat flux.

Source	Sum of Squares	Df	Mean Square	F-Ratio	P-Value
Between Groups	96,009.1	7	13,715.6	17.55	0.000
Within Groups	87,551.5	112	781.71		
Total (Corr.)	183,561	119			

Weight loss caused by thermal degradation of the board surface, which was exposed to radiant heat (Table 2), maintained the same course. Trend lines for the individual board thicknesses have a parallel direction, so it can be assumed that the weight loss is uniform per each unit of thickness (Figure 7). Difference can be seen in the samples with a thickness of 12 mm.

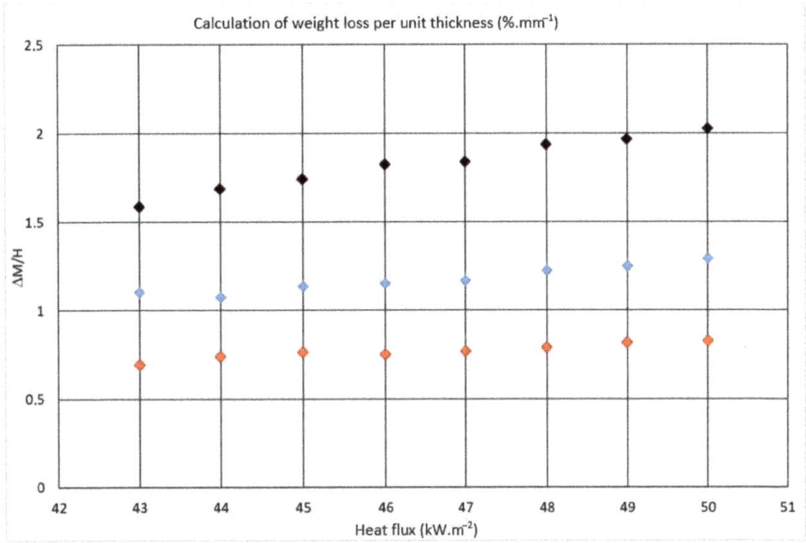

Figure 7. Graphical dependence of the weight loss ΔM and sample thickness H ratio on the heat flux. Legend: black point, 12 mm thickness; blue point, 15 mm thickness; red point, 18 mm thickness.

The results clearly show the following dependence: with increasing thickness, the weight loss decreases, but the ratio of weight loss and sample thickness ΔM/H is rather similar in all heat fluxes (Figure 7). A similar scenario is observed at a thickness of 15 mm. The highest weight loss is observed at a thickness of 12 mm, where an area of light weight of the material burns down and it is possible to observe a slight increase in weight loss with increasing heat flux. There are several studies that confirm the reduction of the weight loss and prolongation of time-to-ignition by appropriate fire-retardant treatment of the boards [3,33,34].

Statistical analysis confirmed this fact. The heat flux (Table 5) and material thickness (Table 6) have no statistically significant effect on the change in sample weight.

Table 5. ANOVA table for Col_3 Weight loss by Col_1 heat flux.

Source	Sum of Squares	Df	Mean Square	F-Ratio	P-Value
Between Groups	5.11703E7	7	7.31005E6	2.05	0.0554
Within Groups	4.00277E8	112	3.5739E6		
Total (Corr.)	4.51448E8	119			

Table 6. ANOVA table for Col_3 Weight loss by Col_2 thickness.

Source	Sum of Squares	Df	Mean Square	F-Ratio	P-Value
Between Groups	5.78793E6	2	2.89396E6	0.76	0.4701
Within Groups	4.4566E8	117	3.80906E6		
Total (Corr.)	4.51448E8	119			

The ANOVA table decomposes the variance of Col_3 Weight loss (%) into two components: a between-group component and a within-group component. The F-ratio, which in this case equals 2.0454, is a ratio of the between-group estimate to the within-group estimate. Since the P-value of the F-test is greater than or equal to 0.05, there is not a statistically significant difference between the mean Col_3 Mass loss (%) from one level of Col_1 (heat flux) to another at the 5% significance level.

Additionally, since the P-value of the F-test is greater than or equal to 0.05, there is not a statistically significant difference between the mean Col_3 Weight loss (%) from one level of Col_2 thickness to another at the 5% significance level.

3.2. Determination of Critical Ignition Temperature

Critical ignition temperature was monitored by means of thermocouples placed below and above the test sample. The increase in temperature was monitored on the upper surface of the 15 mm OSB, which was exposed to a direct radiant heat flux of 44, 46, 48 and 50 kW.m^{-2} for 300 s. The OSB boards with 15 mm thickness are the most commonly used building material in the construction of soffits.

The first set of experiments with a heat flux density of 44 kW.m^{-2} for the period of 300 s resulted in an ignition time of 142 s (Table 7). The temperature at the bottom surface of the board rose minimally (Figure 8). The time to ignition (142 s) was the longest compared to other values obtained at higher heat fluxes (Table 3). The ignition temperatures obtained from the experiments varied. It is not possible to find the correlation with other parameters.

Table 7. Time-to-ignition and ignition temperature at the upper and lower surfaces of a 15 mm thick OSB sample corresponding to the individual radiant heat flux densities.

Heat Flux (kW.m^{-2})	Time to Ignition (°C)	Temperature (°C)	
		Direct Side of Heat	Opposite Side
44	142	287	34
46	70	358	25
48	64	252	26
50	58	319	27

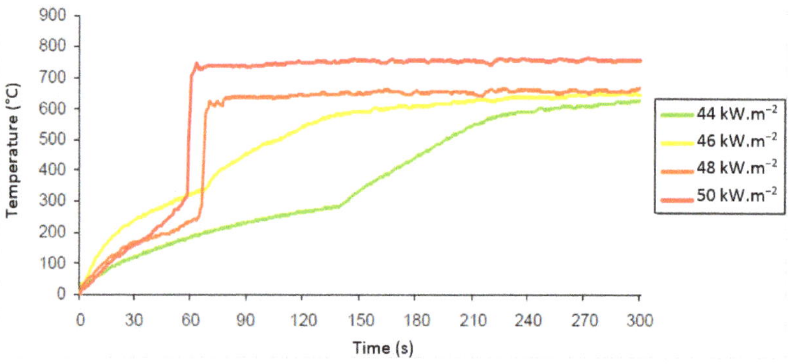

Figure 8. Temperature dependence of the upper surface of an OSB with a thickness of 15 mm on the exposure time to the cone calorimeter.

The temperature at the upper surface of the sample rose sharply to about 700 °C after ignition (Figure 8). The temperature at the bottom surface of the sample rose slowly and had a comparable course in all heat fluxes (Figure 8).

The temperature curves have a smooth linear course and show dependence of the temperature at the upper surface on the time of exposure to radiant heat with heat fluxes of 44 kW.m^{-2} and 46 kW.m^{-2} (Figure 8). After ignition, there was a sharp rise in temperature at the upper surface of the sample (Figure 8). The sharp rise in temperature was likely caused by two effects. The first effect was a higher heat flux from the cone calorimeter to the surface of the tested sample and the second effect was a higher reverse heat radiation from flares [35].

Rantuch et al. [21] divided the OSB combustion process into five phases based on extensive research on OSB (14 mm thickness) using a cone calorimeter with heat fluxes of 20, 30, 40, 50 and 60 kW.m^{-2}. The first phase entails a period prior to the sample ignition. The heat received by the samples is consumed while heating them, but the concentration of gaseous combustible products during thermal degradation is not sufficient to initiate the flame combustion. The second phase occurs after ignition of the sample. The amount of released flammable gases is sufficient to maintain a steady-flame combustion and at the same time a carbonized layer has not yet formed on the surface of the OSB. The mass burning rate as well as the rate of heat release have, therefore, high values. After the formation of a carbonized layer (Figure 4e) on the exposed surface of the samples, a steady burning phase occurs. The rate of heat release is almost constant, and the weight loss is uniform. This is followed by the burning phase of the superheated sample, characterized by significant pyrolysis in its entire volume (Figure 4d).

The increase in the concentration of pyrolysis products is manifested by a more pronounced combustion and, thus, also by an increase in the rate of heat release. After complete burnout of the gaseous pyrolysis products, flameless combustion of the sample occurs with low weight loss [21].

OSB is often a part of the structural elements in the exterior, and therefore it is also crucial to carry out large-scale fire tests [36–38]. Sultan [36] showed that the effect of insulation types on the fire resistance of exterior wall assemblies with OSB sheathing can be considered significant for specific conditions.

4. Conclusions

The ignition process cannot be defined by a single characteristic. Our research monitored important parameters, specifically type of the sample (OSB), thickness, heat flux density, weight loss, ignition temperature and critical ignition temperature, and modified conditions specified in the test procedures. OSBs with a thickness of 12, 15 and 18 mm were analysed with a focus on the critical heat flux and the ignition temperature. The experimentally determined value of the heat flux density was 43 kW.m^{-2}, which represented the critical heat flux. In all cases, dependences of the ignition time on the external heat flux were confirmed and there was a correlation between the ignition time, weight loss and intensity of radiant heat.

The following results were obtained from the conducted experiments:

- As the heat flux density increased, the ignition time decreased in all thicknesses of analysed OSB.
- The ignition time increased with increasing thickness of the OSB, and the weight loss decreased with increasing thickness of the OSB at a constant heat flux.
- With increasing board thickness, the weight loss decreased. The largest average weight loss of 24.31% was recorded in a 12 mm OSB that was exposed to a radiant heat flux of 50 kW.m^{-2}; the lowest average weight loss of 12.44% was recorded in a sample of 18 mm OSB that was exposed to a radiant heat flux of 43 kW.m^{-2}.
- The ignition time is significantly dependent on the thickness of the OSB sample and on the value of the heat flux. As the heat flux increases, the ignition time shortens; as the thickness of the OSB increases, the ignition time extends. The largest weight

loss of 27.22% was recorded in an OSB with a thickness of 12 mm and a density in the range of 500 to 550 kg.m^{-3}, which was exposed to a radiant heat flux of 50 kW.m^{-2}. The lowest weight loss of 11.91% was found in an OSB with a thickness of 18 mm and a density in the range of 550–600 kg.m^{-3}, which was exposed to a radiant heat flux of 43 kW.m^{-2}.

- The weight loss is not significantly dependent on the ignition time and the thickness of the OSB.
- Critical temperature of an OSB with a thickness of 15 mm that was exposed to heat flux densities of 44 kW.m^{-2} and 46 kW.m^{-2} had a linear character; at the heat flux densities of 48 kW.m^{-2} and 50 kW.m^{-2}, it had an initially linear course, but due to ignition, a sharp rise in temperature was noted at the upper surface of the sample. The sharp rise in temperature was caused by two effects. The first effect was a higher heat flux from the cone calorimeter to the surface of the tested sample, and the second effect was a higher reverse heat radiation from flares.

Analysed parameters, such as the time-to-ignition parameter, related weight loss of the OSBs, the density and thickness of the OSBs, radiant heat flux density, distance of the ignition source from the material and determination of the critical ignition temperature with a modified arrangement of the test equipment and horizontal placement of the sample, confirmed the importance and complexity of these parameters for a better understanding of the critical ignition conditions of the OSB, as well as the combustion process.

Author Contributions: Conceptualization, I.T. and M.I.; methodology, I.T.; software, I.M.; validation, I.T., M.I. and J.H.; formal analysis, I.M.; investigation, I.T.; resources, I.T. and M.I.; data curation, M.I.; writing—original draft preparation, I.T.; writing—review and editing, I.M.; project administration, I.M.; funding acquisition, I.T. All authors have read and agreed to the published version of the manuscript.

Funding: This article was supported by the Cultural and Educational Grant Agency of the Ministry of Education, Science, Research and Sport of the Slovak Republic on the basis of the project KEGA 0014UKF-4/2020 Innovative learning e-modules for safety in dual education.

Data Availability Statement: Not applicable for studies not involving humans or animals.

Acknowledgments: This article was supported by the Project KEGA 0014UKF-4/2020 Innovative learning e-modules for safety in dual education.

Conflicts of Interest: The founding sponsors had no role in the design of the study; in the collection, analyses, or interpretation of data; in the writing of the manuscript or in the decision to publish the results.

References

1. Mantanis, G.I.; Athanassiadou, E.T.; Barbu, M.C.; Wijnendaele, K. Adhesive systems used in the European particleboard, MDF and OSB industries. *Wood Mater. Sci. Eng.* **2017**, *13*, 104–116. [CrossRef]
2. Bušterová, M.; Tureková, I.; Martinka, J.; Harangózo, J. The Influence of Heat Flux on the Ignition of OSB Boards. *Spektrum* **2011**, *11*, 5–7. (In Slovak)
3. Gaff, M.; Kacik, F.; Gasparik, M. The effect of synthetic and natural fire-retardants on burning and chemical characteristics of thermally modified teak (*Tectona grandis L. f.*) wood. *Constr. Build. Mater.* **2019**, *200*, 551–558. [CrossRef]
4. Notice of Special Conditions for the Award of a National Eco-Label, Product Group wood-based panels. Available online: https://www.minzp.sk/files/eu/oznamenie_dosky-baze-dreva_2-17.pdf (accessed on 9 March 2017). (In Slovak)
5. CEN Standard EN 300: 2007. *Oriented Strand Boards (OSB). Definitions, Classification and Specifications*; European Committe for Standartion: Brussels, Belgium, 2007.
6. Lunguleasa, A.; Dumitrascu, A.E.; Ciobanu, V.D. Comparative Studies on Two Types of OSB Boards Obtained from Mixed Resinous and Fast-growing Hard Wood. *Appl. Sci.* **2020**, *10*, 6634. [CrossRef]
7. Li, W.; Chen, C.; Shi, J.; Mei, C.; Kibleur, P.; Van Acker, J.; Van den Bulcke, J. Understanding the mechanical performance of OSB in compression tests. *Constr. Build. Mater.* **2020**, *260*, 119837. [CrossRef]
8. Igaz, R.; Krišťák, L.; Ružiak, I.; Gajtanska, M.; Kučerka, M. Thermophysical Properties of OSB Boards versus Equilibrium Moisture Content. *BioResources* **2017**, *12*, 8106–8118.
9. Tudor, E.M.; Dettendorfer, A.; Kain, G.; Barbu, M.C.; Réh, R.; Krišťák, L. Sound-Absorption Coefficient of Bark-Based Insulation Panels. *Polymers* **2020**, *12*, 1012. [CrossRef]

10. Michalovič, R. Fire assesment of different floor materials. In Proceedings of the 19th International Scientific Conference Crisis Management in a Specific Environment, Faculty of Security Engineering, Žilina, Slovakia, 21–22 May 2014; University of Žilina: Žilina, Slovakia, 2014; pp. 497–504. (In Slovak)

11. Štefka, V. *Composite Wood Materials. Part 2: Technology of Agglomerated Materials*, 1st ed.; Technical University in Zvolen: Zvolen, Slovakia, 2005; 205p. (In Slovak)

12. Očkajová, A.; Kučerka, M. *Materials and Technology 1*, 1st ed.; Woodtechnology Monography; Belanium–Matej Bel University: Banská Bystrica, Slovakia, 2011; 115p.

13. SterlingOSB Material Safety Data Sheet. (esi.info). Available online: 10366_1506070430142.pdf (accessed on 1 June 2017).

14. Vandličková, M.; Marková, I.; Makovická Osvaldová, L.; Gašpercová, S.; Svetlík, J.; Vraniak, J. Tropical Wood Dusts—Granulometry, Morfology and Ignition Temperature. *Appl. Sci.* **2020**, *10*, 7608. [CrossRef]

15. Marková, I.; Ladomerský, J.; Hroncová, E.; Mračková, E. Thermal parameters of beech wood dust. *BioResources* **2018**, *13*, 3098–3109. [CrossRef]

16. Makovická Osvaldová, L.; Gašpercová, G. The evaluation of flammability properties regarding testing methods. *Civ. Environ. Eng. Sci. Tech. J.* **2015**, *11*, 142–146. (In Slovak)

17. Lowden, L.A.; Hull, T.R. Flammability behaviour of wood and a review of the methods for its reduction. *Fire Sci. Rev.* **2013**, *2*, 1–19. [CrossRef]

18. Sinha, A.; Nairn, J.A.; Gupta, R. Thermal degradation of bending strenght of plywood and oriented strand board: A kinetics approach. *Wood Sci. Technol.* **2011**, *45*, 315–330. [CrossRef]

19. CEN Standard EN ISO 13943: 2018. *Fire safety. Vocabulary*; European Committe for Standartion: Brussels, Belgium, 2018.

20. *ISO 3261: :1975 Fire tests — Vocabulary*; International Organization for Standardization: Geneva, Switzerland, 1975.

21. Rantuch, P.; Kacíková, D.; Martinka, J.; Balog, K. The Influence of Heat Flux Density on the Thermal Decomposition of OSB. *Acta Fac. Xylologiae Zvolen Publica Slovaca* **2005**, *57*, 125–134.

22. Babrauskas, V. *Ignition of wood: A Review of the State of the Art*; Interscience Communications Ltd.: London, UK, 2001; pp. 71–88.

23. Babrauskas, V. *Ignition Handbook*, 1st ed.; Fire Science Publishers: Issaquah, WA, USA, 2003.

24. *ISO 5657:1997 Reaction to Fire Tests—Ignitability of Building Products using a Radiant Heat Source*; International Organization for Standardization: Geneva, Switzerland, 1997.

25. *CEN Standard EN 323:1993—Wood-Based Panels—Determination of Density*; European Committe for Standartion: Brussels, Belgium, 1993.

26. CEN Standard EN 13501-2: 2018. *Fire Classification of Construction Products and Building Elements. Part 1: Classification using Data from Reaction to Fire Tests*; European Committe for Standartion: Brussels, Belgium, 2018.

27. ASTM E84. *Standard Test Method for Surface Burning Characteristics of Building Materials*; ASTM: West Conshohocken, PA, USA, 2001.

28. Ayrilmis, N.; Kartal, S.; Laufenberg, T.; Winandy, J.; White, R. Physical and mechanical properties and fire, decay, and termite resistance of treated oriented strandboard. *For. Prod. J.* **2005**, *55*, 74.

29. Östman, B.A.L.; Mikkola, E. European classes for the reaction to fire performance of wood products. *Holz als Roh-und Werkst.* **2006**, *64*, 327–337. [CrossRef]

30. Rantuch, P.; Hladová, M.; Martinka, J.; Kobetičová, H. Comparison of Ignition Parameters of Oven-Dried and Non-Dried OSB. *Fire Prot. Saf. Sci. J.* **2019**, *13*, 63–79. (In Slovak)

31. Simms, D.L.; Law, M. The ignition of wet and dry wood by radiation. *Combust. Flame* **1967**, *11*, 377–388. [CrossRef]

32. Mitterová, I.; Garaj, J. The Effect of the Retardation Treatment on the Mass Loss of the Thermally Loaded Spruce Wood and OSB Board. *Fire Prot. Saf. Sci. J.* **2019**, *13*, 51–55. (In Slovak)

33. Osvaldova, L.M.; Gaspercova, S.; Mitrenga, P. The influence of density of test speciemens on the quality assessment of retarding effects of fire retardants. *Wood Res.* **2016**, *61*, 35–42.

34. Osvaldova, L.M.; Osvald, A. Flame retardation of wood. 4th International Conference on Manufacturing Science and Engineering (ICMSE 2013), Dalian, China, 30–31 March 2013. *Adv. Mater. Res.* **2013**, *690–693*, 1331–1334.

35. Bušterová, M. *Effect of Heat Flux on the Ignition of Selected Board Materials*; Faculty of Materials Science and Technology in Trnava, Slovak University of Technology in Bratislava: Bratislava, Slovakia, 2011; 149p.

36. Sultan, M.A. Fire Resistance of Exterior Wall Assemblies for Housing and Small Buildings. *Fire Technol.* **2020**. [CrossRef]

37. Manzello, S.L.; Suzuki, S.; Nii, D. Full-Scale Experimental Investigation to Quantify Building Component Ignition Vulnerability from Mulch Beds Attacked by Firebrand Showers. *Fire Technol.* **2017**, *53*, 535–551. [CrossRef] [PubMed]

38. Koo, E.; Pagni, P.J.; Weise, D.R.; Woycheese, J.P. Firebrands and spotting ignition in large-scale fires. *Int. J. Wildland Fire* **2011**, *19*, 818–843. [CrossRef]

 *polymers*

Article

Non-Isothermal Thermogravimetry of Selected Tropical Woods and Their Degradation under Fire Using Cone Calorimetry

Linda Makovicka Osvaldova [1,*], Ivica Janigova [2] and Jozef Rychly [2]

[1] Department of Fire Engineering, Faculty of Security Engineering, University of Zilina, 01026 Zilina, Slovakia
[2] Polymer Institute, Slovak Academy of Sciences, 84541 Bratislava, Slovakia; ivica.janigova@savba.sk (I.J.); jozef.rychly@savba.sk (J.R.)
* Correspondence: linda.makovicka@fbi.uniza.sk

Abstract: For selected tropical woods (Cumaru, Garapa, Ipe, Kempas, Merbau), a relationship was established between non-isothermal thermogravimetry runs and the wood weight loss under flame during cone calorimetry flammability testing. A correlation was found for the rate constants for decomposition of wood in air at 250 and 300 °C found from thermogravimetry and the total time of sample burning related to the initial mass. Non-isothermal thermogravimetry runs were assumed to be composed from 3 theoretical runs such as decomposition of wood into volatiles itself, oxidation of carbon residue, and the formation of ash. A fitting equation of three processes was proposed and the resulting theoretical lines match experimental lines.

Keywords: tropical wood; non-isothermal thermogravimetry; deconvolution of thermogravimetry runs; cone calorimetry testing; heat-release rate

check for **updates**

Citation: Makovicka Osvaldova, L.; Janigova, I.; Rychly, J. Non-Isothermal Thermogravimetry of Selected Tropical Woods and Their Degradation under Fire Using Cone Calorimetry. *Polymers* **2021**, *13*, 708. https://doi.org/10.3390/polym13050708

Academic Editor: Roman Réh

Received: 18 January 2021
Accepted: 22 February 2021
Published: 26 February 2021

Publisher's Note: MDPI stays neutral with regard to jurisdictional claims in published maps and institutional affiliations.

1. Introduction

Wood is a natural, flammable material, and therefore fire-risk properties represent some obstacle for its applicability in constructions. The fire-technical properties of common woods such as spruce, oak, beech, etc. [1–4] used in different constructions and buildings have been given much attention, while less attention has been paid to tropical woods as they represent more complex systems [5,6]. Differences in fire behavior from that of the classical woods may be the result of the tropical woods' morphology, differences in content of hemicelluloses, cellulose, and lignin, as well as the characteristics and amount of inorganic and organic substances of low molar mass.

Tropical and subtropical trees have been imported, processed, and sold in Central Europe since the end of the 1970s, desirable for their excellent physical and mechanical characteristics and innovative color finish in comparison with European native trees. The good stability and lifespan of some exotic tree species make them the ideal choice for decking balcony and swimming pool flooring, etc., especially in places with increased humidity and high probability of sudden weather changes [7]. Wood from tropical trees is used both in natural and modified forms, such as Thermwood [8,9]. However, as with any wood, tropical woods are a natural combustible material, and their fire properties are worth of further study. Less attention has been paid to tropical woods, regarding their potential fire-risk, than to the more typical woods such as e.g., spruce, oak, beech, etc. [10–13] used in different constructions and buildings. Tropical woods are more complex systems [1,4,5], leading to more difficulty in characterizing their fire properties. Comparison of the combustibility of woods of different origin can be complicated due to differences in density, and therefore different initial masses of samples at uniform dimensions; this disparity may affect the time to ignition. There are also the different admixtures of both organic and inorganic nature with the result that unambiguous ordering of combustibility may be more complex [14–18].

This contribution, a continuation of previous research [19] is an attempt to characterize the thermal stability of 5 tropical woods, three of South American origin (Cumaru, Garapa, Ipe) and two of Southeast Asian origin (Kempas and Merbau), by non-isothermal thermogravimetry in air and oxygen. Simultaneously, the accompanying changes in mass during burning were monitored using cone calorimetry. The burning was characterized by the classical parameters read from cone calorimetry such as MARHE (Maximum average rate of heat emission) and FIGRA (fire growth rate) as well as by the effective heat of combustion and consumption of oxygen.

However, it is worth noting that in addition to its combustion or mechanical resistance properties, the choice of a wood for a particular application must always be conditioned to the careful analysis of its sustainable character regarding its conditions of exploitation and availability of its species to long term. The goal of this research is not therefore to support the extreme deforestation and illegal logging of rainforests, but to do research on the types of trees which might be used in buildings, but to partially examine their response to fire. In the case of some tropical tree species (in natural form or modified), getting to know as much as possible about their fire properties (willingness to ignite and burn) may limit their use in practice, as they might not meet the fire regulations of some countries. By pinpointing these, we can regulate and limit their wider use.

2. Materials and Methods

2.1. Materials

The samples came from a Slovak supplier DLH SLOVAKIA s.r.o. in the form of boards without any surface treatment. Three were from the South American continent—Cumaru, Garapa, Ipe—and two from South-east Asia—Kempas and Merbau. Specification lists of respective woods are in papers [19,20].

Samples were selected carefully regarding their moisture content and density. The samples were selected from the cut boards such that samples' density did not differ by more than $\pm 5\%$ kg/m^3. The content of water in the samples was measured gravimetrically and conditioned to constant weight.

The density of the samples (kg/m^3) at 12% relative humidity was: Cumaru (1070) > Ipe (1050) > Kempas (880 > Merbau (830) > Garapa (790).

Samples of the dimensions $100 \times 100 \times 20$ mm (± 1 mm) were used in cone calorimetry testing. 4 parallel experiments were performed for each wood type.

Figure 1 shows the cross-section and microscopic structure of the selected tree species: (a) Cumaru, (b) Garapa, (c) Ipe, (d) Kempas, (e) Merbau.

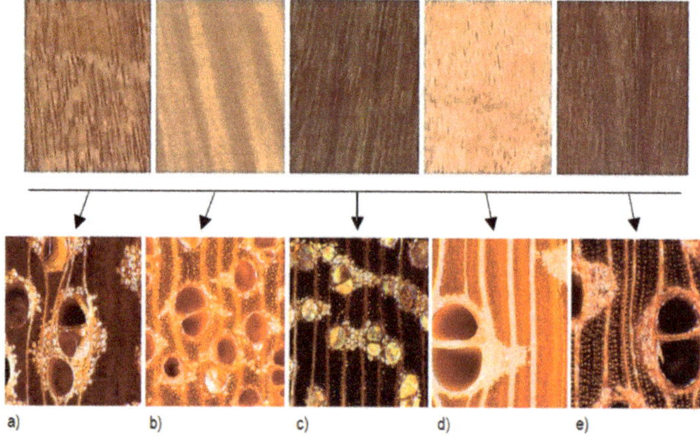

Figure 1. Microscopic structure (cross-section) of tropical tree species from listed as: (**a**) Cumaru, (**b**) Garapa, (**c**) Ipe, (**d**) Kempas, (**e**) Merbau [19].

2.2. Method

2.2.1. Thermogravimetry

The wood board samples were reduced to sawdust; the mass used for the measurements was between 1–2 mg.

The change of the sample mass with increasing temperature was measured using a Mettler–Toledo TGA/SDTA 851^e instrument with a gas flow of 30 mL/min, in a temperature range from room temperature up to 550 °C and a heating rate of 10 °C/min. Temperature calibration was performed with Indium and aluminum standards. The repeatability of the experiments was excellent; the scatter between two parallel measurements in the temperature region of an active decomposition between 300–400 °C was lower than 1 °C. The average line from two parallel measurements was taken [21–23].

2.2.2. Thermogravimetry Reaction Scheme

The formation of volatile degradation products from wood, as observed by non-isothermal thermogravimetry, is a complex process. Before the appearance of molecules capable of escaping the system, wood may undergo a sequence of reaction steps that are essentially based on cleavage of the weakest bonds in the macromolecules of cellulose, hemicellulose, and lignin. This results in smaller fragments which are both volatile and nonvolatile under instantaneous conditions. Simultaneously, depolymerization may occur, as well as crosslinking that may enormously complicate the process of the formation of volatiles [20,21,24].

In this paper, we introduce Scheme 1 of wood W decomposition into volatiles V, where S are the products remaining on the reaction pan.

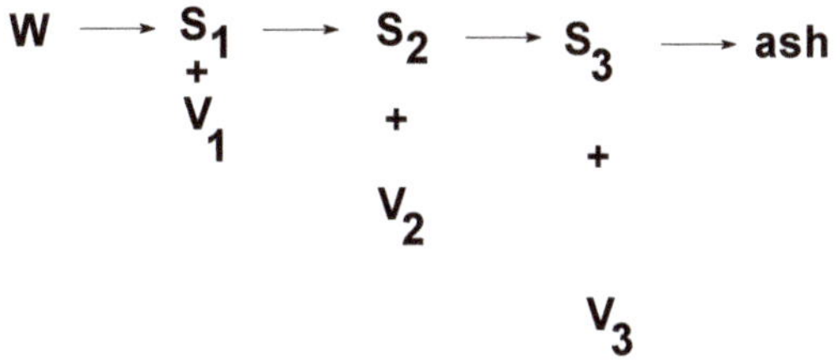

$$W \longrightarrow S_1 \longrightarrow S_2 \longrightarrow S_3 \longrightarrow ash$$
$$+ \qquad + \qquad +$$
$$V_1 \qquad V_2 \qquad V_3$$

Scheme 1. Wood decomposition as the sequence of component processes.

If the temperature program is appropriately chosen so that respective steps of Scheme 1 become well separated, then from the viewpoint of volatiles formed it may well be replaced by the Scheme 2 that we have used for the deconvolution of observed experimental lines [21].

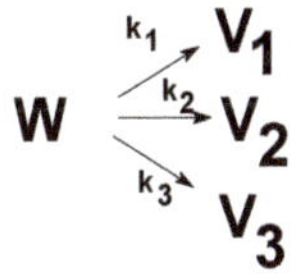

$$W \begin{array}{c} \xrightarrow{k_1} V_1 \\ \xrightarrow{k_2} V_2 \\ \xrightarrow{k_3} V_3 \end{array}$$

Scheme 2. Simplified Scheme 1.

The complex non-isothermal thermogravimetry curve may then be described as being composed of three independent processes each being described by the first-order scheme, i.e.,

$$-\frac{dm}{dt} = km \tag{1}$$

In a non-isothermal mode,

$$-\frac{dm}{dT}\frac{dT}{dt} = A\,exp\left(-\frac{E}{RT}\right)m$$

or alternatively,

$$-\frac{dm}{m} = \frac{A}{\beta}\exp\left(-\frac{E}{RT}\right)dT \tag{2}$$

where m—actual mass, T—temperature in Kelvins, R—universal gas constant, A and E—pre-exponential factor and activation energy and is the linear heating rate.

$$\beta = \frac{dT}{dt}$$

After integration of this equation, we obtain:

$$m = m_0 \exp\left[-\frac{A}{\beta}\int_{T_0}^{T}\exp\left(-\frac{E}{RT}\right)dT\right] \tag{3}$$

For the process composed of 3 temperature dependent components—"waves", we have:

$$m = m_0 \sum_{i=1}^{j}\alpha_i\exp\left[-\frac{A_i}{\beta}\int_{T_0}^{T}\exp\left(-\frac{E_i}{RT}\right)dT\right] \tag{4}$$

Provided that mass changes are expressed as a percentage of the initial mass, m_0, the parameters α_i, A_i, E_i, where α_i is a fraction of respective component process, may be found by a nonlinear regression analysis (Levenberg, Marquardt algorithm) applied to curves of the experimental mass m vs. temperature T, from the initial temperature T_0 to a final temperature T of the experiment. The rate constant, k_i, corresponding to a given temperature is expressed as $k_i = A_i\exp(-E_i/RT)$.

2.2.3. Cone Calorimetry

The cone calorimeter used was product of Fire Testing Laboratory Ltd. UK. The peak release rate (PkHRR) from a cone calorimeter, a key factor in fire assessment, is measured by monitoring the oxygen consumption based on the difference of air flows between the cone entrance and the cone extraction pipe of the cone calorimeter system. Calculations can then be executed based on empirical knowledge of the combustion of macromolecules, namely that an average of 13.1 MJ of heat is released per one kilogram of consumed O_2 [25]. The heat flow was set at 35 kW/m^2.

Other readouts were ignition time, total burning time, weight loss, effective heat consumption (EHC), total O_2 consumption.

Each sample of the given dimensions was placed into the steel frame in horizontal position. The only part of the sample exposed to heat was its surface, not its edges. To prevent the material from peeling and the degradation products from dripping, the sample was wrapped in aluminum foil on the bottom and the sides. The radiator, which has a shape of a truncated cone and provided a uniform source of radiation, was placed above the sample. The temperature was regulated using 3 thermocouples and a thermostat. Weight measurements were carried out using the load cell of a tensometer with a readability of 0.01 g. A 10 kV spark generator equipped with a safety shut-off mechanism was used to ignite the volatiles from the test specimens [26]. There was a steady flow of air of 24 l/h throughout the burning channel.

3. Results

3.1. Non-Isothermal Thermogravimetry

Figure 2 demonstrates the effect of atmosphere (air and oxygen) on the mass loss of Cumaru wood when heated at a rate of 10 °C/min. The mass loss was precipitated by a slight increase in the mass, probably due to peroxidation. The main mass loss then is faster in oxygen than in air.

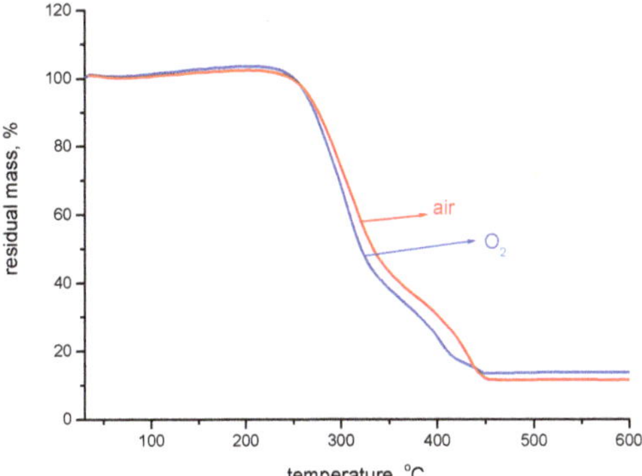

Figure 2. The average non-isothermal thermogravimetry–temperature runs (from 2 parallel experiments) in oxygen, and air for Cumaru wood at the rate of heating 10 °C/min.

For deconvolution of experimental runs and for evaluation of corresponding pre-exponential factor and activation energy of the main fraction of volatiles formed from wood the function 1) and $j = 3$ was used.

An example of such a deconvolution for Garapa samples in air is shown in Figure 3.

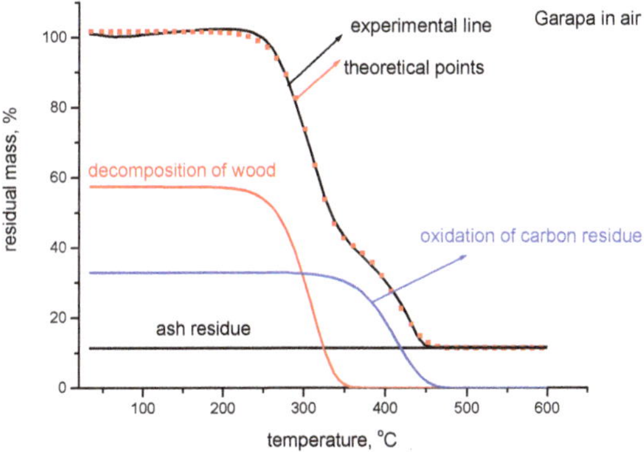

Figure 3. The deconvolution of non-isothermal thermogravimetry–temperature run into corresponding component processes with the use of Equation (1) for Garapa wood in air at the rate of heating of 10 °C/min. Points represent the theoretical run.

Table 1 gives the formal summary of parameters extracted from average experimental runs-mass of the sample vs. temperature according to Equation (4). The main component of the wood degradation is depicted in red. From this set of data, the rate constants of the wood decomposition were determined for 250 and 300 °C (Table 2). The decomposition of wood itself (red line), oxidation of carbon residue (blue line) and slight temperature changes of the ash (black line) can be seen.

Table 1. Set of parameters from deconvolution of non-isothermal thermogravimetry runs (Equation (4)) for samples of tropical woods measured in air and oxygen (the rate of heating of 10 °C/min.) (α_i is the fraction ascribed to the respective component of the whole degradation process, i = 1,2 and 3).

Sample	Gas	α_1	A_1/β, (°C)$^{-1}$	E_1, kJ/mol	α_2	A_2/β, (°C)$^{-1}$	E_2, kJ/mol	α_3	A_3/β, (°C)$^{-1}$	E_3, kJ/mol
Cumaru	air	0.460	2.2×10^5	97.4	0.465	8.6×10^7	106.1	0.060	7.2×10^{-6}	26.9
Garapa	air	0.110	28.2	166.3	0.574	6.6×10^7	103.3	0.328	5.1×10^7	121.5
Ipe	air	0.311	1.1×10^8	133.4	0.602	2.2×10^7	101.3	0.110	2.1×10^{-8}	21.1
Kempas	air	0.302	6.7×10^3	78.6	0.590	2.2×10^7	97.6	0.124	3.2×10^{-7}	37.0
Merbau	air	0.329	2.3×10^6	113.5	0.604	3.6×10^5	81.0	0.076	8.4×10^{-7}	32.7
Cumaru	oxygen	0.438	8.2×10^4	88.8	0.485	1.2×10^9	116.6	0.081	4.0×10^{-7}	31.3
Garapa	oxygen	0.136	19.6	160.3	0.554	2.6×10^9	118.4	0.334	4.0×10^5	41.2
Ipe	oxygen	0.337	0.095	104.5	0.572	5.6×10^8	113.8	0.095	8.9×10^{-4}	0.004
Kempas	oxygen	0.341	3.2×10^3	71.4	0.596	4.5×10^7	98.8	0.080	5.2×10^{-7}	40.4
Merbau	oxygen	0.359	1.9×10^5	95.2	0.599	5.8×10^6	92.2	0.040	1.4×10^{-8}	12.1

Table 2. Set of parameters extracted from deconvolution of non-isothermal thermogravimetry runs (the Equation (4), Table 1) for samples of tropical woods measured in air and oxygen decomposed with the rate of heating of 10 °C/min. The above set of data is related to second component process (the red data) which was used for determination of the rate constants of second component of wood mass loss at 250 and 300 °C.

Sample	Gas	α_2	A_2/β, (°C)$^{-1}$	E_2, kJ/mol	$k_2300 \times 10^3$, s^{-1}	$k_2250 \times 10^4$, s^{-1}
Cumaru	air	0.465	8.6×10^7	106.1	3.6	4.3
Garapa	air	0.574	6.6×10^7	103.3	4.9	6.2
Ipe	air	0.602	2.2×10^7	101.3	2.5	3.3
Kempas	air	0.590	2.2×10^7	97.6	5.3	7.6
Merbau	air	0.604	3.6×10^5	81.0	2.8	5.5
Cumaru	oxygen	0.485	1.2×10^9	116.6	5.5	5.4
Garapa	oxygen	0.554	2.6×10^9	118.4	8.3	7.8
Ipe	oxygen	0.572	5.6×10^8	113.8	4.6	4.8
Kempas	oxygen	0.596	4.5×10^7	98.8	8.5	11.9
Merbau	oxygen	0.599	5.8×10^6	92.2	4.3	6.9

3.2. Cone Calorimetry

The cone calorimetry runs were recorded by the averaging 4 parallel experiments with sample Cumaru (Figure 4); all important parameters for respective samples are in Table 3. In contribution [19] we stated that repeatability of respective experiments was high. Typical heat-release rate (HRR)–time runs involve a sharp increase of the heat released just after ignition. After that time, the flame is spread along the surface and, in its close proximity, the flame then penetrates through the layer of carbonaceous residue underneath.

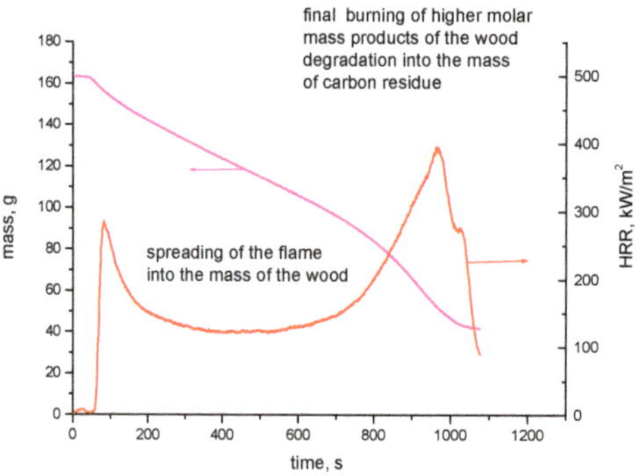

Figure 4. The average HRR–time and mass-loss runs (from 4 parallel experiments) at 35 kW/m^2 for Cumaru.

Table 3. Parameters read from cone calorimetry measurements of tropical wood samples C (Cumaru), G (Garapa), K (Kempas), M (Merbau) and I (Ipe). Average from 4 parallel experiments.

Sample Average Line from 4 Parallel Runs	m Initial (g)	m_{lost} (g)	Time to Ignition/Time of Burning (s)	EHC (MJ/kg)	Peak HRR (kW/m^2)	TOC (g)	MARHE (kW/m^2)	FIGRA kW/m^2/s
Cumaru	163.3	121.5	$45 \pm 4/1067 \pm 30$	15.6	457.2 ± 42.3	115.9	176.5 ± 5.1	0.457
Garapa	180.0	118.5	$90 \pm 10/1439 \pm 49$	13.6	275.0 ± 28.2	98.4	107.4 ± 6.3	0.215
Kempas	139.9	112.7	$37 \pm 2/1200 \pm 34$	17.1	307.8 ± 9.8	115.8	155.6 ± 7.4	0.296
Merbau	253.3	183.0	$91 \pm 16/1676 \pm 50$	17.4	$438.4 \pm 32,8$	194.8	187.9 ± 8.4	0.291
Ipe	227.0	171.9	$89 \pm 5/1340 \pm 64$	16.9	500.9 ± 17.9	177.6	224.6 ± 8.6	0.444

Where m_{lost}—total loss of mass during burning; time to ignition/time of burning–time to ignition is time interval from insertion of sample bellow cone heater to ignition (piloted ignition by spark)—time of burning is time of flame burning of sample until self-extinction; EHC—effective heat of combustion, Peak HRR—maximum heat-release rate corresponding to the second maximum of record; TOC—total oxygen consumed during combustion; MARHE—a maximum value of ARHE = $(\int HRR^*t^* \, dt)/(t_1)$, integral in nominator is from time of ignition to t_1—time of burning; FIGRA—the ratio of the peak HRR/time corresponding to this maximum.

It was assumed that the high-molar-mass products of the degradation of the wood's degradation were pushed into the bottom of the carbonaceous layer by the heat front, and partially accumulate there. In the final stage of the wood burning, a second maximum appears, representing the burning of the accumulated products, followed by the extinction of the flame. The HRR–time runs are, in fact, a first derivative of mass-loss runs, as may be also seen on Figure 4. Table 3 presents the time of ignition, time of burning, HRR peak, MARHE, and FIGRA and other averaged parameters.

When looking at Table 3, it is difficult to decide which of the five samples was the most flammable. However, when testing it by Bunsen burner shoved that Ipe and Merbau samples continued burning even when the burner was removed from direct contact with the sample, while Garapa, Kempas, and Cumaru extinguished. This alights with the values of MARHE parameters related to the second maximum, as well as with FIGRA (Figure 5). It is generally valid that the higher MARHE and/or FIGRA, the more flammable sample is.

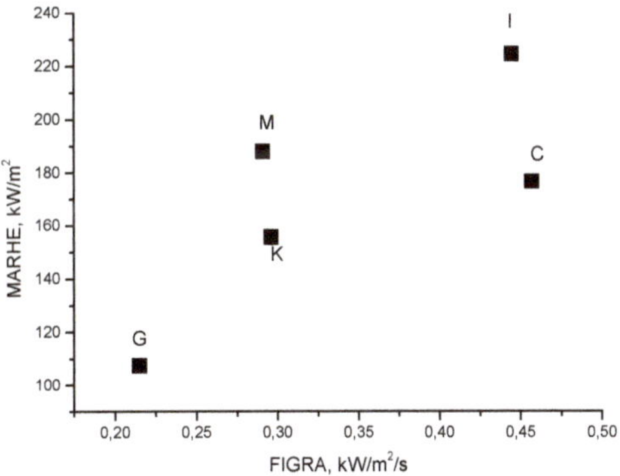

Figure 5. Correlation between MARHE and FIGRA parameters (average values) for respective samples of tropical woods: C—Cumaru, G—Garapa, I—Ipe, K—Kempas, M—Merbau.

Figure 6 shows the average plots of mass loss related to 100% of the original mass during the burning of the woods. In the region of almost-linear decay, designated by arrows, quasi-isothermal conditions may be expected to occur on the surface of the burning material.

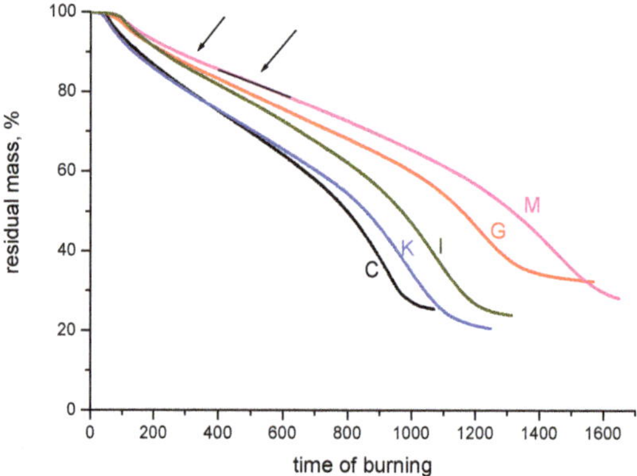

Figure 6. The average mass loss–time runs (from 4 parallel experiments) at 35 kW/m^2 for C—Cumaru, G—Garapa, I—Ipe, K—Kempas, M—Merbau.

The density of the wood, and thus the initial mass at identical dimensions, is one of the factors determining the time to ignition (Figure 7). For samples of the same quality, (spruce) this is quite evident, while tropical trees differing also in structure and the amount of other additional compounds show larger scatter of respective experiments. Kempas, Cumaru, and Garapa approximately follow the straight line, but Ipe and Merbau decline. Ipe and Merbau samples have lower time to ignition than expected from the linear plot of Figure 7, probably because they are more ignitable than other woods.

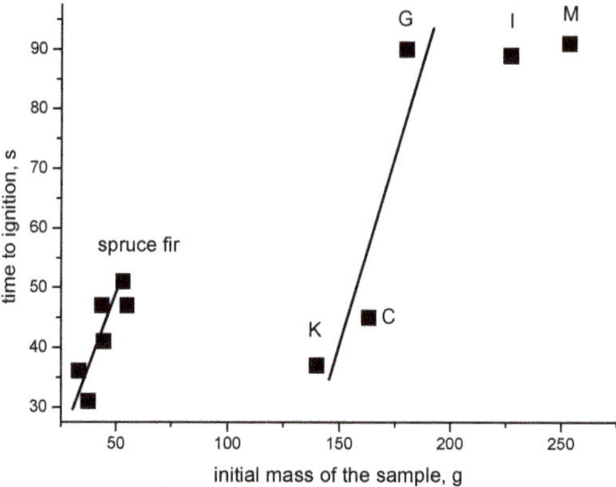

Figure 7. Time to ignition in dependence on initial mass. The cone radiancy was 35 kW/m^2. The times of ignition of spruce were presented in the paper [19], G—Garapa, C—Cumaru, K—Kempas, I—Ipe and M—Merbau.

Time to ignition appears also to be related to the total amount of oxygen consumed in burning (Figure 8), as if the ignition requires more oxygen to successfully initiate the burning. A quite convincing demonstration of a possible link between thermogravimetry runs and cone calorimetry burning may be seen in Figure 9, where rate constants from non-isothermal thermogravimetry determined for 250 and °C correlate quite well with the total time of burning related to the initial mass of samples.

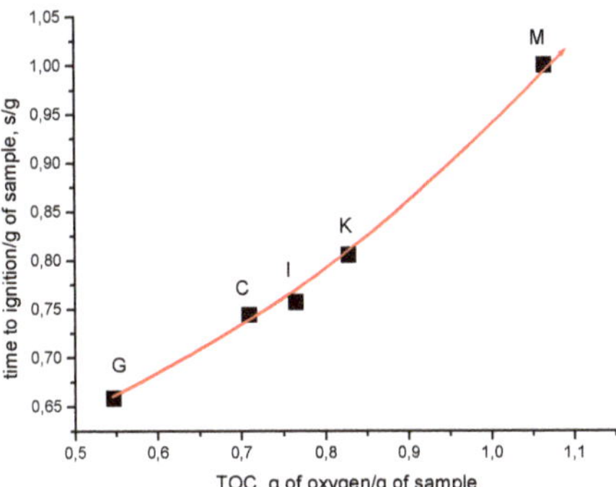

Figure 8. Correlation between time to ignition and the total oxygen consumed (TOC) related both to the initial mass of the sample for C—Cumaru, G—Garapa, I—Ipe, K—Kempas, M—Merbau.

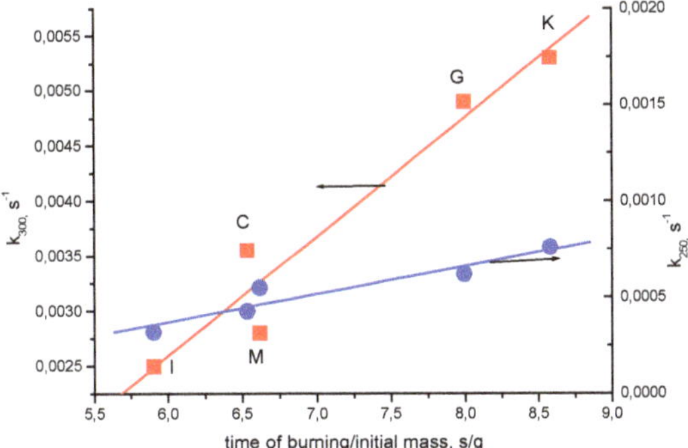

Figure 9. The rate constants at 250 and 300 °C determined from non-isothermal thermogravimetry vs. the total time of burning related to the initial mass of wood C—Cumaru, G—Garapa, I—Ipe, K—Kempas, M—Merbau.

Figure 10 shows the relationship between EHC and total oxygen consumed (TOC) divided by the mass lost during burning. The straight line produced in both cases of tropical woods, as well as spruce fir, has the shape: EHC = 1.008 + 16.1 × TOC/mass lost where the slope about 16 kJ/g is higher than parameter 13.1 kJ/g of consumed oxygen implemented in cone calorimeter. It may be of interest that the above correlation is also valid for spruce samples that were thinner than tropical woods [19].

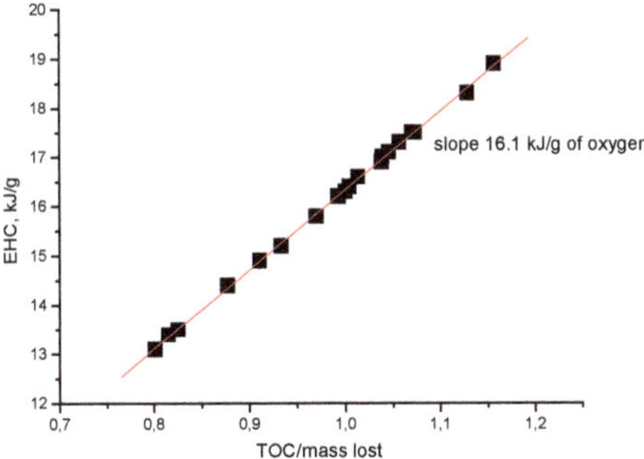

Figure 10. Universal correlation between EHC and total oxygen consumed related to unity of mass lost during burning (all parallel experiments involved).

4. Discussion

A correlation has been found for the rate constants for the decomposition of selected tropical woods in air at 250 and 300 °C, and the total time of sample burning related to the initial mass. The sequence of rate constants of wood degradation was as follows: Kempas > Garapa > Merbau > Cumaru > Ipe.

Decomposition of wood in thermogravimetry works with small initial mass of the sample, and thus with the direct release of volatile decomposition products into gaseous phase. However, in a cone calorimetry burning with much higher initial mass, the back effect of the heat from cone heater on degradation products formed during burning may keep them beneath the upper carbonaceous layer [27–32].

As expected, the time to ignition was affected by the initial mass of sample. This was valid for Cumaru, Ipe, and Kempas while Garapa and Merbau decline. This is probably due the fact that the latter two woods give the decomposition phenomenon composed formally from two processes [19,32–36].

The sequence of MARHE parameter related to the second maximum of the heat-release rate–time cone calorimetry runs is different from the sequence obtained from thermogravimetry. The order of flammability of examined woods is: Ipe (238) > Merbau (188) > Cumaru (176) > Kempas (155) > Garapa (107), given in kW/m^2. This sequence correlates quite well with FIGRA parameters. At the same time, this may describe the relative tendency of the burning wood to retain some less volatile degradation products below the upper carbonaceous layer.

A linear correlation between EHC and TOC related to a mass lost during burning has been confirmed. The slope is however higher (about 16 kJ/g of consumed oxygen) than that implemented into cone calorimeter (13.1 kJ/g of consumed oxygen), which indicates that the stoichiometry of burning involves a significant portion of products such as carbon in the form of solid residue and soot, as well as carbon monoxide.

5. Conclusions

The degradation of tropical woods was evaluated in air and oxygen at a heating rate of 10 °C/min, while cone calorimetry tests were performed in air with a cone radiance of $35\ kW/m^2$. The experimental conditions of decomposition of wood using thermogravimetry mean that considerably lower initial masses of a sample are used and the direct release of volatile decomposition products into the gaseous phase occurs. The sequence of the rate constants of wood degradation was as follows: Kempas > Garapa > Merbau > Cumaru > Ipe. The same decaying sequence was found for flammability of samples expressed as a total time of their burning related to the initial mass.

The larger initial mass of the samples in cone calorimetry burning created the conditions for the back effect of the heat from the cone heater contributing temporary retention of higher-molar-mass degradation products burning beneath the upper carbonaceous layer, which is demonstrated by the appearance of the second maximum on curves of heat-release rate vs. time. This maximum is brought about by their subsequent release.

Author Contributions: Conceptualization, L.M.O.; methodology, J.R.; formal analysis, I.J.; investigation, L.M.O.; resources, I.J.; writing—original draft preparation, L.M.O. and J.R.; writing—review and editing, L.M.O.; visualization, I.J.; supervision, L.M.O. All authors have read and agreed to the published version of the manuscript.

Funding: This research received no external funding.

Institutional Review Board Statement: Not applicable.

Informed Consent Statement: Not applicable.

Data Availability Statement: Data sharing not applicable.

Conflicts of Interest: The authors declare no conflict of interest.

References

1. Momoh, M.; Horrocks, A.R.; Eboatu, A.N.; Kolawole, E.G. Flammability of tropical woods—I. Investigation of the burning parameters. *Polym. Degrad. Stab.* **1996**, *54*, 403–411. [CrossRef]
2. Elvira-Leon, J.C.; Chimenos, J.M.; Isabal, C.; Monton, J.; Formosa, J.; Haurie, L. Epsomite as flame retardant treatment for wood: Preliminary study. *Constr. Build. Mater.* **2016**, *126*, 936–942. [CrossRef]
3. Wiemann, M.C. Characteristics and availability of commercially important woods. In *Wood Handbook: Wood as an Engineering Material*; Forest Products: Madison, WI, USA, 2010; pp. 2.1–2.45.
4. Zelinka, S.L.; Passarini, L.; Matt, F.J.; Kirker, G.T. Corrosiveness of Thermally Modified Wood. *Forests* **2020**, *11*, 50. [CrossRef]
5. Koklukaya, O.; Carosio, F.; Grunlan, J.C.; Wagberg, L. Flame-retardant paper from wood fibers functionalized via layer-by-layer assembly. *ACS Appl. Mater. Interfaces* **2015**, *7*, 23750–23759. [CrossRef] [PubMed]
6. Östman, B.A.-L. Fire performance of wood products and timber structure. *Int. Wood Prod. J.* **2017**, *8*, 74–79. [CrossRef]
7. Tsapko, Y.; Tsapko, O.; Bondarenko, O. Determination of the laws of thermal resistance of wood in application of fire-retardant fabric coatings. *East. Eur. J. Enterp. Technol.* **2020**, *2*, 104. [CrossRef]
8. Sun, L.; Wu, Q.; Xie, Y.; Cueto, R.; Lee, S.; Wang, Q. Thermal degradation and flammability behavior of fire-retarded wood flour/polypropylene composites. *J. Fire Sci.* **2016**, *34*, 226–239. [CrossRef]
9. Sebio-Puñal, T.; Naya, S.; López-Beceiro, J.; Tarrio-Saavedra, J.; Artiaga, R. Thermogravimetric analysis of wood, holocellulose, and lignin from five wood species. *J. Therm. Anal. Calorim.* **2012**, *109*, 1163–1167. [CrossRef]
10. Jiang, J.; Li, J.; Hu, J.; Fan, F. E_ect of nitrogen phosphorus flame retardants on thermal degradation of wood. *Constr. Build. Mater.* **2010**, *24*, 2633–2637. [CrossRef]
11. Ecochard, Y.; Decostanzi, M.; Negrell, C.; Sonnier, R.; Caillol, S. Cardanol and eugenol based flame retardant epoxy monomers for thermostable networks. *Molecules* **2019**, *24*, 1818. [CrossRef]
12. Belykh, S.; Novoselova, J.; Novoselov, D. Fire Retardant Coating for Wood Using Resource-Saving Technologies. In *International Scientific Conference Energy Management of Municipal Facilities and Sustainable Energy Technologies EMMFT 2018*; Murgul, V., Pasetti, M., Eds.; EMMFT-2018; Advances in Intelligent Systems and Computing; Springer: Cham, Switzerland, 2020; Volume 982. [CrossRef]
13. Maqsood, M.; Langensiepen, F.; Seide, G. The efficiency of biobased carbonization agent and intumescent flame retardant on flame retardancy of biopolymer composites and investigation of their melt-spinnability. *Molecules* **2019**, *24*, 1513. [CrossRef]
14. Kasymov, D.; Agafontsev, M.; Perminov, V.; Martynov, P.; Reyno, V.; Loboda, E. Experimental Investigation of the Effect of Heat Flux on the Fire Behavior of Engineered Wood Samples. *Fire* **2020**, *3*, 61. [CrossRef]
15. Lin, C.; Karlsson, O.; Mantanis, G.I.; Jones, D.; Sandberg, D. Fire Retardancy and Leaching Resistance of Pine Wood Impregnated with Melamine Formaldehyde Resin in-Situ with Guanyl-Urea Phosphate/Boric Acid. In *Wood & Fire Safety*; Makovicka Osvaldova, L., Markert, F., Zelinka, S., Eds.; WFS 2020; Springer: Cham, Switzerland, 2020; pp. 83–89. [CrossRef]
16. Gaugler, M.; Grigsby, W.J. Thermal Degradation of Condensed Tannins from Radiata Pine Bark. *J. Wood Chem. Technol.* **2009**, *29*, 305–321. [CrossRef]
17. Poletto, M. Thermal degradation and morphological aspects of four wood species used in lumber industry. *Rev. Arvore* **2016**, *40*, 941–948. [CrossRef]
18. Kang, S.; Kwon, M.; Choi, J.Y.; Choi, S. Thermal Boundaries in Cone Calorimetry Testing. *Coatings* **2019**, *9*, 629. [CrossRef]
19. Makovicka Osvaldova, L.; Kadlicova, P.; Rychly, J. Fire Characteristics of Selected Tropical Woods without and with Fire Retardant. *Coatings* **2020**, *10*, 527. [CrossRef]
20. Gerad, J.; Guibal, D.; Paradis, S.; Cerre, J.C. *Tropical Timber Atlas Technological Characteristics and Uses*, 1st ed.; Editions Quæ: Versailles Cedex, France, 2017; pp. 600–602.
21. Thermal Analysis. Available online: https://chem.libretexts.org/Bookshelves/Analytical_Chemistry/Book%3A_Physical_Methods_in_Chemistry_and_Nano_Science_(Barron)/02%3A_Physical_and_Thermal_Analysis/2.08%3A_Thermal_Analysis#Thermogravimetric_Analysis_(TGA).2FDifferential_Scanning_Calorimetry_(DSC) (accessed on 22 December 2020).
22. Ergun, B.; Ilyas, D.; Turkay, T.; Hilmi, T. Thermal analysis of oriental beech sawdust treated with some commercial wood preservatives. *Maderas. Cienc. Tecnol.* **2017**, *19*, 329–338. [CrossRef]
23. Mustafa, B.G.; Mat Kiah, M.H.; Andrews, G.E.; Phylaktou, H.N.; Li, H. Toxic Gas Emissions from Plywood Fires. In *Wood & Fire Safety*; Makovicka Osvaldova, L., Markert, F., Zelinka, S., Eds.; WFS 2020; Springer: Cham, Switzerland, 2020; pp. 50–57. [CrossRef]
24. Hasburgh, L.E.; Stone, D.S.; Zelinka, S.L.; Plaza, N.Z. Characterization of Wood Chemical Changes Caused by Pyrolysis During Flaming Combustion Using X-ray Photoelectron Spectroscopy. In *Wood & Fire Safety*; Makovicka Osvaldova, L., Markert, F., Zelinka, S., Eds.; WFS 2020; Springer: Cham, Switzerland, 2020; pp. 22–27. [CrossRef]
25. *ASTM E1354-17 Standard Test Method for Heat and Visible Smoke Release Rates for Materials and Products Using an Oxygen Consumption Calorimeter*; ASTM International: West Conshohocken, PA, USA, 2017. [CrossRef]
26. Sonnier, R.; Otazaghine, B.; Ferry, L.; Lopez-Cuesta, J.M. Study of the combustion e_ency of polymers using a pyrolysis-combustion flow calorimeter. *Combust. Flame* **2013**, *160*, 2182–2193. [CrossRef]
27. Xu, Q.; Chen, L.; Harries, K.A.; Zhang, F.; Liu, Q.; Feng, J. Combustion and charring properties of five common constructional wood species from cone calorimeter tests. *Constr. Build. Mater.* **2015**, *96*, 416–427. [CrossRef]

28. Giraldo, M.P.; Haurie, L.; Sotomayor, J.; Lacasta, A.M.; Monton, J.; Palumbo, M.; Nazzaro, A. Characterization of the fire behaviour of tropical wood species for use in the construction industry. In Proceedings of the WCTE 2016: World Conference on Timber Engineering, Vienna, Austria, 22–25 August 2016; Technischen Universität Graz: Graz, Austria, 2016; pp. 5387–5395.

29. Tsapko, Y.; Lomaha, V.; Bondarenko, O.P.; Sukhanevych, M. Research of Mechanism of Fire Protection with Wood Lacquer. *Mater. Sci. Forum* **2020**, *1006*, 32–40. [CrossRef]

30. Kamiya, K.; Sugawa, O. Odor and FT-IR Analysis of Chemical Species from Wood Materials in Pre-combustion Condition. In *Wood & Fire Safety*; Makovicka Osvaldova, L., Markert, F., Zelinka, S., Eds.; WFS 2020; Springer: Cham, Switzerland, 2020; pp. 35–40. [CrossRef]

31. Mazela, B.; Batista, A.; Grześkowiak, W. Expandable Graphite as a Fire Retardant for Cellulosic Materials—A Review. *Forests* **2020**, *11*, 755. [CrossRef]

32. Neykov, N.; Antov, P.; Savov, V. Circular Economy Opportunities for Economic Efficiency Improvement in Wood-based Panel Industry. In Proceedings of the 11th International Scientific Conference "Business and Management 2020", Vilnius, Lithuania, 7–8 May 2020. [CrossRef]

33. Tudor, E.M.; Scheriau, C.; Barbu, M.C.; Réh, R.; Krišťák, Ľ.; Schnabel, T. Enhanced Resistance to Fire of the Bark-Based Panels Bonded with Clay. *Appl. Sci.* **2020**, *10*, 5594. [CrossRef]

34. Rezvani Ghomi, E.; Khosravi, F.; Mossayebi, Z.; Saedi Ardahaei, A.; Morshedi Dehaghi, F.; Khorasani, M.; Neisiany, R.E.; Das, O.; Marani, A.; Mensah, R.A.; et al. The Flame Retardancy of Polyethylene Composites: From Fundamental Concepts to Nanocomposites. *Molecules* **2020**, *25*, 5157. [CrossRef] [PubMed]

35. Mascarenhas, A.R.; Sccoti, M.S.; de Melo, R.R.; de Oliveira Corrêa, F.L.; de Souza, E.F.; Pimenta, A.S. Quality assessment of teak (Tectona grandis) wood from trees grown in a multi-stratified agroforestry system established in an Amazon rainforest area. *Holzforschung* **2020**, 000010151520200082. [CrossRef]

36. Rensink, S.; Sailer, M.F.; Bulthuis, R.J.H.; Oostra, M.A.R. Application of a Bio-Based Coating for Wood as a Construction Material: Fire Retardancy and Impact on Performance Characteristics. In *Wood & Fire Safety*; Makovicka Osvaldova, L., Markert, F., Zelinka, S., Eds.; WFS 2020; Springer: Cham, Switzerland, 2020; pp. 90–96. [CrossRef]

Article

Changes of Meranti, Padauk, and Merbau Wood Lignin during the ThermoWood Process

Danica Kačíková [1], Ivan Kubovský [1,*], Milan Gaff [2] and František Kačík [1]

1   Faculty of Wood Sciences and Technology, Technical University in Zvolen, T.G. Masaryka 24, 960 01 Zvolen, Slovakia; kacikova@tuzvo.sk (D.K.); kacik@tuzvo.sk (F.K.)

2   Faculty of Forestry and Wood Sciences, Czech University of Life Sciences in Prague, Kamýcká 129, 165 00 Praha 6-Suchdol, Czech Republic; gaff@fld.czu.cz

*   Correspondence: kubovsky@tuzvo.sk

Abstract: Thermal modification is an environmentally friendly process in which technological properties of wood are modified using thermal energy without adding chemicals, the result of which is a value-added product. Wood samples of three tropical wood species (meranti, padauk, and merbau) were thermally treated according to the ThermoWood process at various temperatures (160, 180, 210 °C) and changes in isolated lignin were evaluated by nitrobenzene oxidation (NBO), Fourier-transform infrared spectroscopy (FTIR), and size exclusion chromatography (SEC). New data on the lignins of the investigated wood species were obtained, e.g., syringyl to guaiacyl ratio values (S/G) were 1.21, 1.70, and 3.09, and molecular weights were approx. 8600, 4300, and 8300 $g \cdot mol^{-1}$ for meranti, padauk, and merbau, respectively. Higher temperatures cause a decrease of methoxyls and an increase in C=O groups. Simultaneous degradation and condensation reactions in lignin occur during thermal treatment, the latter prevailing at higher temperatures.

Keywords: meranti; padauk; merbau; thermal treatment; wood lignin

Citation: Kačíková, D.; Kubovský, I.; Gaff, M.; Kačík, F. Changes of Meranti, Padauk, and Merbau Wood Lignin during the ThermoWood Process. *Polymers* **2021**, *13*, 993. https://doi.org/10.3390/polym13070993

Academic Editor: Antonios N. Papadopoulos

Received: 2 March 2021
Accepted: 18 March 2021
Published: 24 March 2021

Publisher's Note: MDPI stays neutral with regard to jurisdictional claims in published maps and institutional affiliations.

1. Introduction

As a renewable composite material, wood is an ideal building material that is easy to work with and offers advantages such as a high strength-to-weight ratio and lower processing energy. However, dimensional instability is one of its major drawbacks, especially for structural uses. Apart from that, it is also susceptible to fungal degradation and to weathering [1–3]. These shortcomings can be minimized by making hydrophilic wood hydrophobic, and thermal treatment is a widely used modification method for this purpose. Some wood products are popular in dark and/or red colour. Wood colour becomes darker and redder with increasing temperature due to formation of chromophores predominantly in lignin [4]. Tropical woods have high commercial value on the market and wood industry due to their good appearance and excellent physical, mechanical, and machinability properties. Despite the immense tree diversity of tropical forests in Brazil, only a few species are well known, explored, and sold in local markets, but many others can provide wood with good properties and high potential for applications in the wood industry [5–7]. Nevertheless, it is desirable to improve some of their properties, e.g., durability, stability, pest resistance, and colour uniformity. Thermal treatment appears to be an environmentally friendly and economical technology to improve the properties and colour of wood products. Thermal treatment is suitable for wood since it is non-toxic and does not require the use of chemicals. The thermal modification of wood at temperatures from 180 to 260 °C leads to hemicellulose and lignin degradation. This process changes the chemical composition of the wood and reduces its hygroscopicity. Thus, thermally modified wood tends to be more dimensionally stable than unmodified wood of the same species [8–11].

One of the main wood components is lignin, the largest source of aromatics on earth, as wood-derived biomass consists of up to 35% of lignin. Lignin is an amorphous cross-linked biopolymer that, in combination with cellulose and hemicelluloses, confers structural stability to plants [12,13]. Many statements about lignin being "energetically utilized" are confessions that come disguised as proud claims, but we still do not know how to utilize lignin on a large scale better than burning it. The vast amounts of technical lignins generated annually by the global pulp and paper industries are still awaiting viable ideas for large-scale and general utilization. In addition, lignin can be used in a broad range of composite materials and serves as a raw material to produce many chemicals [14–16]. However, a large amount of lignin is also produced in other wood processing industries, such as in the recycling of wood products, which also includes thermally modified wood, and this work can contribute to its better utilization. Thermally modified wood has been used to improve wood composite properties, e.g., wood–plastic composites, particleboard, etc. [17].

Although heat-treated wood exhibits some improved properties [18], e.g., hydrophobicity, dimensional stability, decay resistance, and darker colour, however, some mechanical properties deteriorate due to the heat treatment [19]. The bending strength properties for keruing (*Dipterocarpus* spp.) and light red meranti (*Shorea* spp.) wood were affected by heat exposure. Both modulus of elasticity (MOE) and modulus of rupture (MOR) values for both thermally treated wood species increased when subjected to temperatures of 150, 170, and 190 °C, except for 210 °C [20]. The lightness of teak and meranti wood was the most affected colour attribute during thermal treatment [21].

African padauk wood is a versatile material, and with declining rosewood resources, the value of this wood species is increasing. Padauk wood is used for making furniture, musical instruments, and for construction purposes. Although many applications are implemented, it is still possible to obtain many new applications of this material [22,23]. Thermal treatment of iroko and padauk wood caused a decrease in its density, colour darkening, and considerable improvement of dimensional stability [24]. Although merbau wood is often used in outdoor applications, its disadvantages include easy leachability of water-soluble substances that stain the surrounding materials. This shortcoming can be solved by thermal treatment [25,26]. Merbau extractives have potential as an impregnating material for low-quality timber to improve its properties [27,28]. The effect of higher temperatures on selected fire safety features of tropical wood shows that the thermal treatment of merbau and meranti wood significantly increased its flammability and accelerated its combustion [29].

Despite intensive research, there is still a lack of data about chemical changes in properties of many tropical woods during their thermal modification. Therefore, the purpose of this study was to provide more detailed information about the effect of thermal modification on lignin changes in meranti, padauk, and merbau wood species.

2. Materials and Methods

Light red meranti (*Shorea* spp.), padauk (*Pterocarpus soyauxii* Taub.), and merbau (*Intsia* spp.) wood species with dimensions of $20 \times 20 \times 300$ mm (tangential × radial × longitudinal) were thermally modified at various temperatures [30]. Samples were labelled as 20 (untreated), 160, 180, and 210 (according to the applied temperatures), disintegrated to sawdust, and extracted according to the ASTM Standard Test Method [31]. The acid-insoluble lignin (known as Klason lignin) was determined according to the National Renewable Energy Laboratory (NREL) procedure [32]. Procedures for lignin isolation and conditions of its nitrobenzene oxidation (NBO) were reported recently [33]. The molecular weight distribution (MWD) evaluation of dioxane lignins was performed by the previously described method [33]. Fourier transform infrared spectroscopy (FTIR) of isolated dioxane lignin was performed on a Nicolet iS10 spectrometer with Smart iTR ATR accessory. The spectra were collected in an absorbance mode between 4000 and 650 cm^{-1} by accumulating 32 scans at a resolution of 4 cm^{-1} using diamond crystal. All analyses were performed in four replicates. Statistical analysis was performed by applying a one-way analysis of variance, using the probability theory and Fisher's F-test.

3. Results and Discussion

3.1. Changes in Lignin Yields

Lignin is the most stable wood component at thermal treatment. Its yield usually increases during all kinds of biomass pre-treatments accomplished at a low pH and high temperature conditions such as dilute acid pre-treatment, hot water pre-treatment, steam explosion, and at high temperatures. The relative percentage of acid-insoluble lignin (Klason lignin) is higher in the modified material than in that of the untreated ones. This phenomenon is due to formation of pseudo-lignin due to the condensation reactions of degradation products of lignin and polysaccharides [25,34,35]. The results in Table 1 show the different behaviour of meranti, padauk, and merbau wood Klason lignins when subjected to thermal modification. The increase of meranti lignin is relatively high at a temperature of 160 °C, then it decreases unexpectedly. This yield drop can be due to preferential lignin degradation reactions, and it corresponds with the decline of lignin molecular weight determined by size exclusion chromatography (Table 4). In contrast, an increase in lignin was observed after the superheated steam treatment of light red meranti wood [36]. On the other hand, yields of padauk and merbau lignins continuously increase similarly to the increase found during the thermal treatment of various wood species [34,36,37]. We found a higher Klason lignin content (32.42%) with a S/G ratio of 1.21 (Table 2), with differences probably due to different wood species, the location of their origin, etc. Similar results for the S/G ratio were found by Syafii, 2001 [38] for several tropical wood species. He reported that the syringyl to guaiacyl ratio of albizia, gmelina, kapur, and yellow meranti woods are 2.03, 2.02, 1.87, and 1.30, respectively. This means that the lignin structure of the above-mentioned woods is predominated by syringyl units. The content of Klason lignin in yellow meranti was 30.00% [38]. The extraction of lignin from wood by dioxane provides good yields with minimal structural changes, which is why dioxane lignin is often used for structural studies [39,40]. Yields of dioxane lignin increased from 1.7 to 1.8-fold for all three species at a temperature of 210 °C in comparison to the original samples (Table 1). The amount of extracted lignin may depend on the degree of its condensation, but a comparison of the yields of dioxane lignin (Table 1), NBO products (Tables 2–4), and its molecular weights (Tables 5–7) shows that other factors also influence its yields, e.g., functional groups, crosslinking, etc. These factors result in simultaneous degradation and condensation reactions in lignin during the thermal treatment.

Table 1. Klason (KL) and dioxane lignin (DL) yields from untreated and thermally treated meranti, padauk, and merbau (mean ± SD, % odw).

Product	20 °C	160 °C	180 °C	210 °C
Meranti-KL	32.42 ± 0.08	36.67 ± 0.10	36.29 ± 0.09	35.33 ± 0.13
Padauk-KL	33.77 ± 0.10	34.84 ± 0.04	35.53 ± 0.03	39.73 ± 0.09
Merbau-KL	33.75 ± 0.23	33.23 ± 0.07	35.75 ± 0.20	44.61 ± 0.23
Meranti-DL	7.57 ± 0.10	6.79 ± 0.11	7.32 ± 0.06	13.54 ± 0.11
Padauk-DL	9.02 ± 0.16	10.72 ± 0.41	14.20 ± 0.25	16.62 ± 0.31
Merbau-DL	9.01 ± 0.11	7.97 ± 0.12	13.69 ± 0.28	15.37 ± 0.11

The lignin content in meranti (*Shorea almon*) is 23.10% with the ratio of syringyl to guaiacyl (S/G) 0.94 [41].

3.2. Changes in NBO Products

The increasing amount of NBO products indicates preferential degradation reactions, which is also confirmed by the drop of molecular weight determined by the SEC analysis (Table 4, Figure 1). The increase of S/G ratio at higher temperatures suggests that G-type lignin is more prone to the condensation reactions [42].

Table 2. Phenolic aldehydes and acids from nitrobenzene oxidation of meranti dioxane lignins (mean ± SD, %).

Product	20 °C	160 °C	180 °C	210 °C
p-Hydroxybenzoic acid	0.04 ± 0.00	0.05 ± 0.00	0.04 ± 0.01	0.03 ± 0.01
p-Hydroxybenzaldehyde	1.93 ± 0.12	1.44 ± 0.41	1.93 ± 0.12	0.98 ± 0.16
Vanillic acid	0.33 ± 0.00	0.46 ± 0.03	0.65 ± 0.07	0.16 ± 0.03
Vanilline	14.71 ± 0.25	14.19 ± 0.45	15.67 ± 0.09	16.66 ± 0.20
Syringic acid	0.39 ± 0.03	0.38 ± 0.08	0.32 ± 0.00	0.42 ± 0.02
Syringaldehyde	17.77 ± 1.01	18.13 ±0.50	20.26 ± 1.01	22.21 ± 1.76
Total yield on DL	35.17 ± 1.41	34.64 ± 0.52	38.87 ± 1.10	40.47 ± 1.75
S/G ratio	1.21 ± 0.05	1.26 ± 0.08	1.26 ± 0.06	1.35 ± 0.09

Table 3. Phenolic aldehydes and acids from nitrobenzene oxidation of padauk dioxane lignins (mean ± SD, %).

Product	20 °C	160 °C	180 °C	210 °C
p-Hydroxybenzoic acid	0.04 ± 0.01	0.05 ± 0.00	0.04 ± 0.00	0.05 ± 0.00
p-Hydroxybenzaldehyde	3.19 ± 0.12	3.17 ± 0.33	3.08 ± 0.04	3.17 ± 0.16
Vanillic acid	1.09 ± 0.05	1.04 ± 0.16	0.98 ± 0.01	0.92 ± 0.05
Vanilline	7.61 ± 0.07	8.67 ± 0.22	7.96 ± 2.52	3.89 ± 0.07
Syringic acid	1.49 ± 0.06	1.43 ± 0.20	1.41 ± 0.02	1.47 ± 0.09
Syringaldehyde	13.33 ± 1.26	15.28 ± 1.51	14.22 ± 0.50	7.11 ± 0.00
Total yield on DL	26.76 ± 1.56	29.64 ± 1.97	27.68 ± 2.96	16.61 ± 0.24
S/G ratio	1.70 ± 0.13	1.72 ± 0.19	1.75 ± 0.45	1.78 ± 0.02

Table 4. Phenolic aldehydes and acids from nitrobenzene oxidation of merbau dioxane lignins (mean ± SD, %).

Product	20 °C	160 °C	180 °C	210 °C
p-Hydroxybenzoic acid	0.02 ± 0.00	0.04 ± 0.00	0.02 ± 0.00	0.01 ± 0.00
p-Hydroxybenzaldehyde	2.86 ± 0.02	3.26 ±0.16	3.46 ± 0.44	3.23 ± 0.12
Vanillic acid	3.49 ± 0.19	3.57 ± 0.17	4.49 ± 0.98	6.13 ± 0.19
Vanilline	1.18 ± 0.08	2.24 ± 0.02	1.59 ± 0.05	0.30 ± 0.02
Syringic acid	1.24 ± 0.05	1.41 ± 0.11	1.52 ± 0.29	1.44 ± 0.03
Syringaldehyde	13.19 ± 0.46	26.50 ± 0.48	14.18 ± 0.85	6.00 ± 0.45
Total yield on DL	21.98 ± 0.30	37.02 ± 0.05	25.26 ± 0.18	17.11 ±0.59
S/G ratio	3.09 ± 0.16	4.80 ± 0.18	2.58 ± 0.66	1.16 ± 0.04

Table 5. Molecular weights and polydispersity index of lignin from meranti wood (mean ± SD).

Temperature (°C)	M_w (g/mol)	M_n (g/mol)	M_z (g/mol)	PDI
20	8627 ± 256	2549 ± 26	41,157 ± 3769	3.38 ± 0.09
160	8595 ± 122	2470 ± 22	48,388 ± 2696	3.48 ± 0.07
180	5607 ± 109	2217 ± 17	15,651 ± 1432	2.53 ± 0.03
210	4253 ± 75	1761 ± 16	11,071 ± 763	2.42 ± 0.04

Note: M_w: Weight average molecular weight (MW); M_n: Number average MW; M_z: Z average MW; PDI (polydispersity index): M_w/M_n.

Table 6. Molecular weights and polydispersity index of lignin from padauk wood (mean ± SD).

Temperature (°C)	M_w	M_n	M_z	PDI
20	4265 ± 303	1771 ± 32	10,341 ± 529	2.41 ± 0.22
160	4301 ± 103	1850 ± 14	10,237 ± 279	2.32 ± 0.02
180	4504 ± 105	1698 ± 18	12,972 ± 494	2.65 ± 0.07
210	4999 ± 48	1898 ± 12	16,264 ± 669	2.63 ± 0.04

Note: See Table 5 for symbols.

Table 7. Molecular weights and polydispersity index of lignin from merbau wood (mean ± SD).

Temperature (°C)	M_w	M_n	M_z	PDI
20	8284 ± 757	2772 ± 283	37,826 ± 1526	2.99 ± 0.17
160	7461 ± 72	3128 ± 31	20,268 ± 787	2.39 ± 0.02
180	14,567 ± 107	2806 ± 46	113,672 ± 8141	5.19 ± 0.13
210	14,728 ± 729	2815 ± 31	114,080 ± 13,700	5.23 ± 0.29

Note: See Table 5 for symbols.

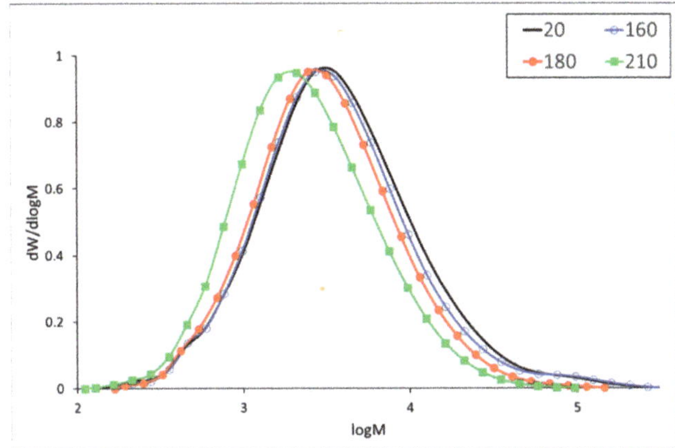

Figure 1. Molecular weight distribution of meranti wood lignins.

Similarly, to meranti, padauk also has SG lignin, containing more syringyl units compared to guaiacyl ones with an S/G ratio of 1.70, and this ratio slightly increases with the temperature (Table 3). No data of S/G ratio in padauk lignin were found in the literature.

An unexpectedly high S/G ratio was determined in merbau lignin (Table 4), indicating a much lower guaiacyl unit content in comparison to meranti and padauk lignin. The high S/G ratio means that this lignin presents a more open matrix, with a lower degree of C-C bonds at the C5-ring position [43].

3.3. Changes of Macromolecular Traits in Lignins

Dioxane lignin isolated from meranti wood before thermal treatment shows typical monomodal molecular weight distribution and a small peak with high molecular weight (Figure 1). Competitive degradation and condensation reactions in lignin can be observed at a temperature of 160 °C, resulting in an increase in the high molecular weight peak (M_z increases from 41,157 to 48,388 g/mol) and partial degradation of the main peak. At higher temperatures, degradation reactions predominate and both the molecular weights and the polydispersity index decrease, e.g., the value of M_w drops to half at a temperature of 210 °C (Table 4, Figure 1). It should be noted that no data on the molecular weight of lignin have been found in scientific literature for either meranti or padauk.

Padauk wood lignin reacts to elevated temperatures differently compared to meranti lignin. From a molecular weight of 4265 g/mol in the case of untreated wood, it increases up to a value of 4999 g/mol at a temperature of 210 °C (Table 5). From molecular weight distribution curves, it is evident that both degradation and condensation reactions occur simultaneously, however, the latter are predominant (Figure 2). The SEC analysis similarly revealed concurrent degradation and condensation reactions by heating poplar lignin [42]. The thermal degradation of lignin causes the formation of phenoxy radicals causing a coupling reaction to form new 4−O−5 and 5−5′ linkages, respectively [44,45]. The predominant condensation reactions result in an increase in molecular weight by almost 20%

when compared to the unmodified sample. An increase in lignin molecular weight was observed in the thermal treatment of spruce wood up to a temperature of 240 °C [46].

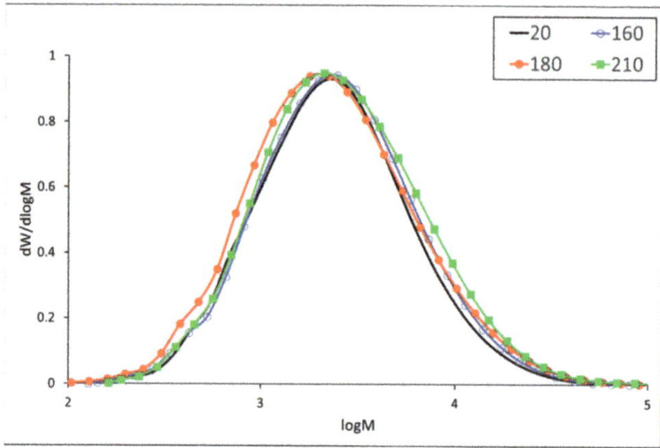

Figure 2. Molecular weight distribution of padauk wood lignin.

The molecular weight of merbau lignin slightly decreases at a temperature of 160 °C (Table 7) and increases dramatically at higher temperatures (by 77%) compared to the untreated sample. Molecular weight distribution curves (Figure 3) show a decrease in both high and low molecular fractions indicating a simultaneous process of degradation and condensation reactions at a lower temperature of treatment. Higher temperatures (180 and 210 °C) predominantly cause the condensation of lignin macromolecules resulting in both higher molecular weights and polydispersity. Similar changes in lignin were found during the thermal treatment of spruce and teak wood, respectively [33,46]. The crosslinking of lignin leading to the increase of molecular weight was observed by [47] during thermoplastic processing of grass lignin with ethylene and vinyl acetate copolymer (EVA).

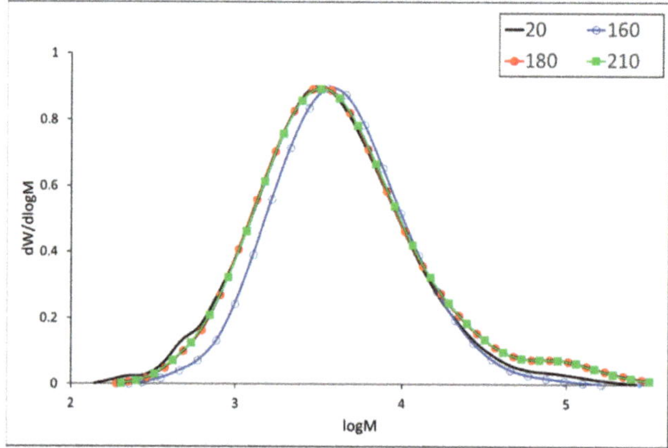

Figure 3. Molecular weight distribution of merbau wood lignin.

3.4. Changes in FTIR Spectra

Chemical changes in thermally treated wood lignin are expressed as the differential FTIR spectral absorbances of selected related bands (Figures 4–6). Their values were obtained as the differences between absorbances after treatment at the given temperature and absorbances of untreated samples (e.g., "ME-Diff 160" means absorbances of meranti lignin treated at 160 °C minus absorbances of untreated meranti lignin). The positive values refer to an increase and the negative values to a decrease in the absorbance. Given the similarity of the trends of the measured absorbances for meranti, padauk, and merbau, the FTIR spectra will be evaluated together (in the case of more significant unequal behavior between them, this will be mentioned). A broad band around 3400 cm^{-1} (OH stretching vibration in alcohols, acids, phenols, and weakly bounded absorbed water in lignin) [22,48] and bands around 2940 and 2840 cm^{-1} (C−H stretching in methyl and methylene groups) [49,50] show only negligible changes in absorbance during the thermal treatment. The absorbance of the band at 1710 cm^{-1} (C=O vibrations of non-conjugated carbonyl groups) increases with the increasing temperature (with a maximum at 210 °C). The increase in intensity may be a result of cleavage of β−O−4 linkage accompanied by the formation of C=O groups formed in lignin during oxidation [51,52]. This trend was observed and explained by several authors as a consequence of increasing the amount of acetyl and carbonyl groups from lignin [53,54]. The absorption intensities near 1600 and 1500 cm^{-1} (C=C stretching vibrations in aromatic structure of lignin) are related to the skeletal vibration of the syringyl and guaiacyl structures in lignin [55,56]. In our case, the band around 1600 cm^{-1} slightly decreased in the meranti and padauk samples treated at lower temperatures (160 and 180 °C) (Figures 1 and 2). On the contrary, it rises at the highest temperature (210 °C), where the increase is higher in padauk (Figures 1 and 2). In the merbau sample, an increase in the band can be observed at each temperature (Figure 3). The band at 1508 cm^{-1} has a different trend in the studied samples. While the meranti and merbau show a gradual increase when subjected to thermal treatment, a permanent decrease can be observed with padauk. An increase in absorbance has been observed by some researchers [57,58], while other authors obtained opposite results [26,52]. The found differences may be due to the different proportions of lignin in the investigated wood species [59]. Absorbance at 1460 cm^{-1} (C−H asymmetric bending in CH$_3$ groups in lignin) relating to hemicellulose and methyl groups in lignin [60], shows a variable trend, which is essentially identical as the reported dependence absorbances under thermal treatment at 1593 cm^{-1}. In samples treated at lower temperatures (160 and 180 °C), it decreases (Figures 1 and 2). On the contrary, it rises slightly at the highest temperature (210 °C), probably due to the condensation reactions in the lignin structure (Figures 1 and 2). The thermal treatment of lignin causes gradual removal of water and methanol to form conjugated ethylene bonds [61]. A permanent regress at 1420 cm^{-1} (aromatic ring vibration in lignin combined with C−H deformation in carbohydrates) [50] was measured (except for merbau, where the band increased at 210 °C). Decreases in absorbances on bands at 1460 and 1420 cm^{-1} indicate the cleavage of methoxyl groups during thermal treatment leading to gradual demethoxylation [52,62]. The bands at 1325 cm^{-1} (C−O vibration in syringyl plus guaiacyl derivatives is characteristic for condensed structures in lignin) [56,63,64], 1265 and 1215 cm^{-1} (C=O stretching and breathing of guaiacyl ring in lignin) [64–66] show a slight decrease at the lowest temperature. This decrease may be due to the cleavage of the ether bond in the lignin structure leading to the elimination of methoxy groups [67]. However, as the temperature rises, the trend reverses and there is a significant increase in these bands (especially in thermal treatment at a temperature of 210 °C). The gradual increase of these bands indicates the development of condensation reactions in lignin [63] (with a significant contribution of guaiacyl). The absorbance value at 1120 cm^{-1} (C−H vibration of syringyl units in lignin) [40] also decreased. The ratio of the relative absorbances of the bands at 1265 and 1120 cm^{-1} shows that a higher temperature induces condensation reactions more easily in guaiacyl units than in syringyl units [68,69]. A decrease of band intensity at 1028 cm^{-1} (C−O stretching) [63] indicate the cleavage of β-alkyl-aryl ether and methoxyl bonds. A similar trend was observed in teak and oak wood [54,57]. These results indicate a significant degradation of the lignin structure, which is most pronounced especially in condensation reactions at the highest temperature value.

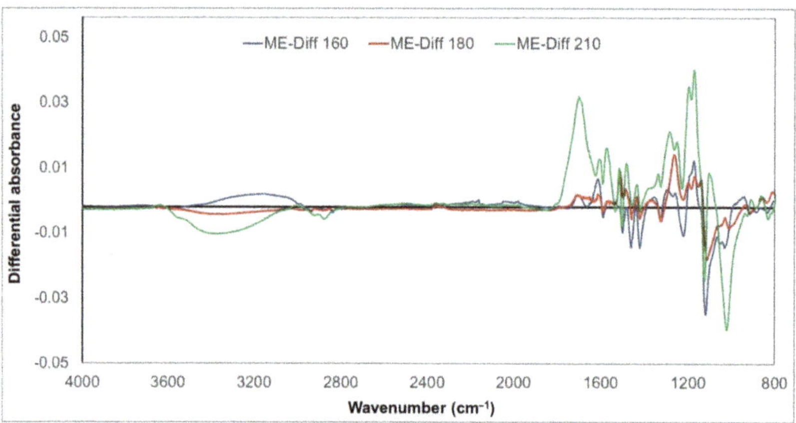

Figure 4. Differential FTIR spectra of thermally treated meranti wood lignin.

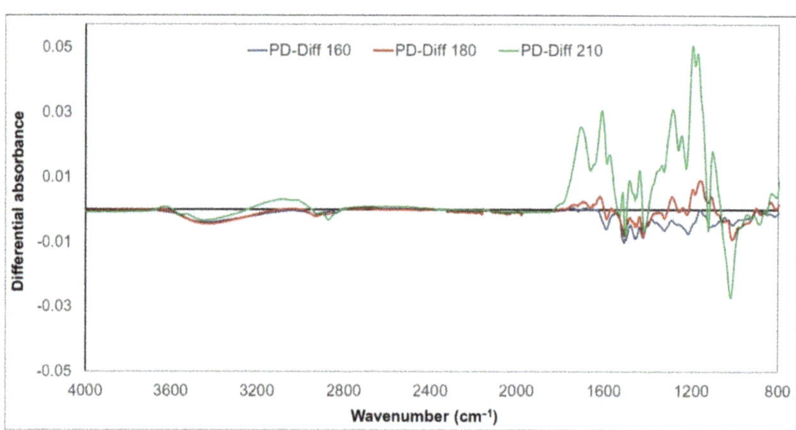

Figure 5. Differential FTIR spectra of thermally treated padauk wood lignin.

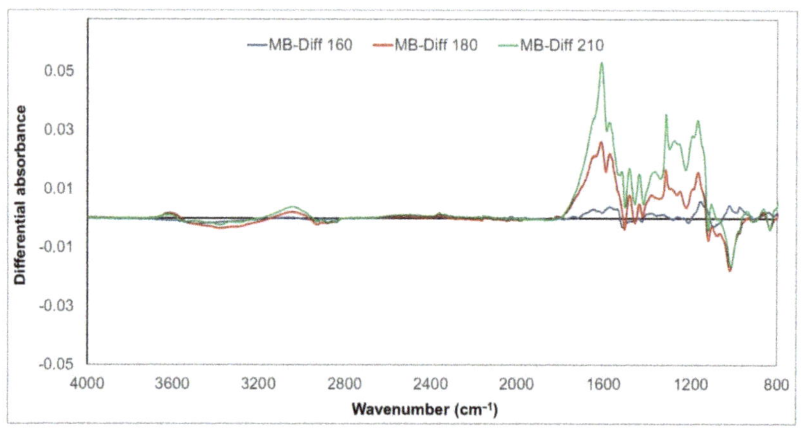

Figure 6. Differential FTIR spectra of thermally treated merbau wood lignin.

3.5. Statistical Evaluation Analysis

The effect of the interaction of thermal modification temperature (20, 160, 180, and 210 °C) and wood species (meranti, padauk, merbau) was evaluated by applying a one-way analysis of variance, using the probability theory and Fisher's F-test. Due to the large range of results, we do not present the tables in this article. Based on the values of the level of significance "P" and Fisher's F-test, these results showed that the effect of the observed interaction (thermal modification temperature and wood species) on the values of M_n, M_w, M_z, S/G, and YIELD has a very significant effect in all monitored cases (Figures 7–12).

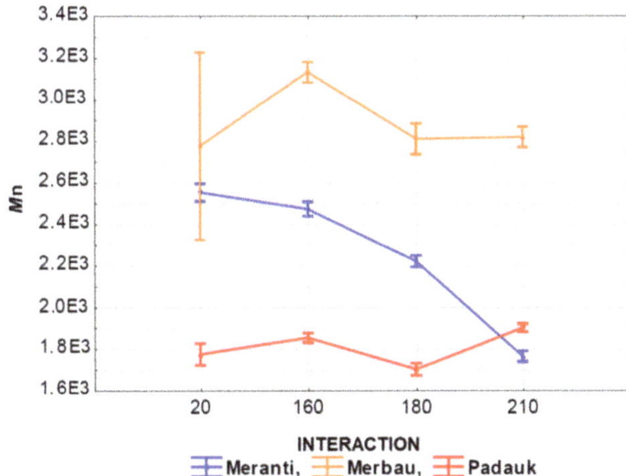

Figure 7. The effect of thermal modification temperature and wood species on number average molecular weight (M_n) values.

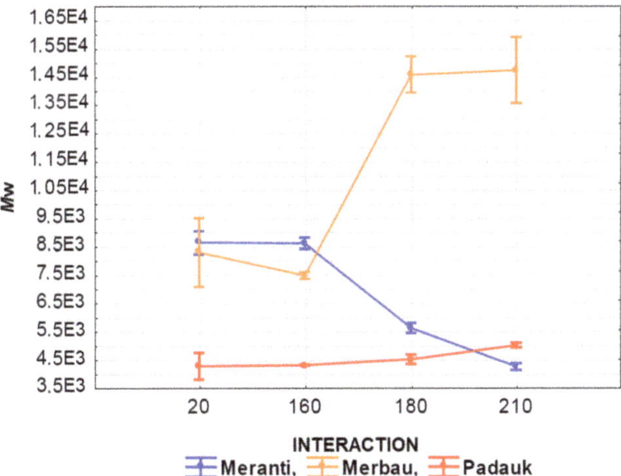

Figure 8. The effect of thermal modification temperature and wood species on weight average molecular weight (M_w) values.

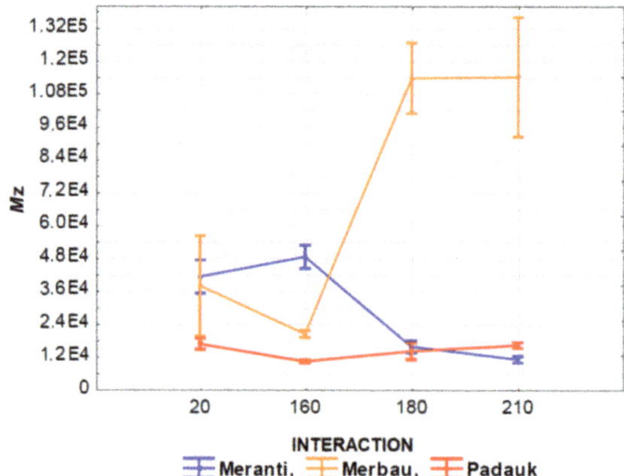

Figure 9. The effect of thermal modification temperature and wood species on Z average molecular weight (M_z) values.

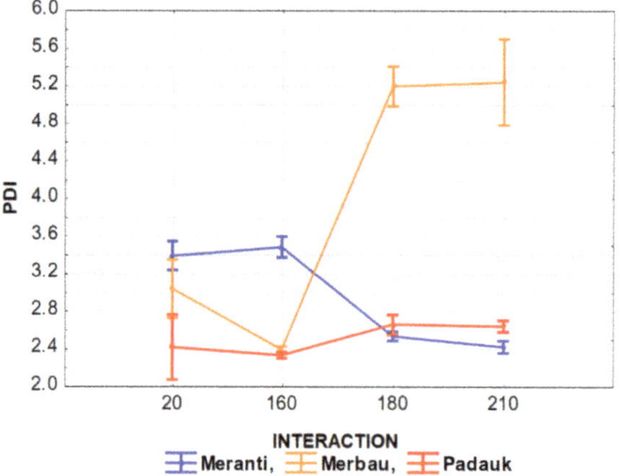

Figure 10. The effect of thermal modification temperature and wood species on polydispersity index.

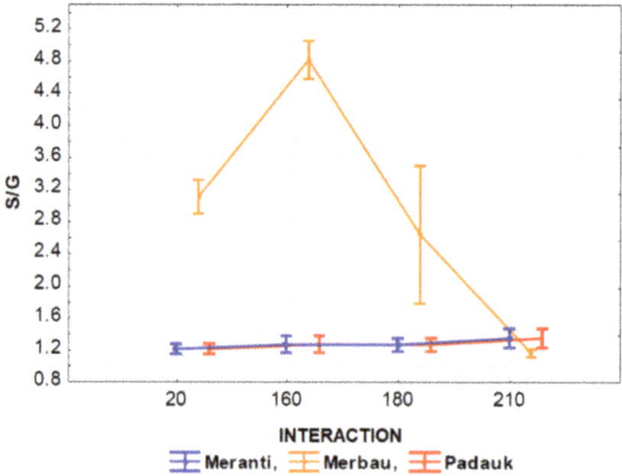

Figure 11. The effect of thermal modification temperature and wood species on syringyl to guaiacyl ratio (S/G) values.

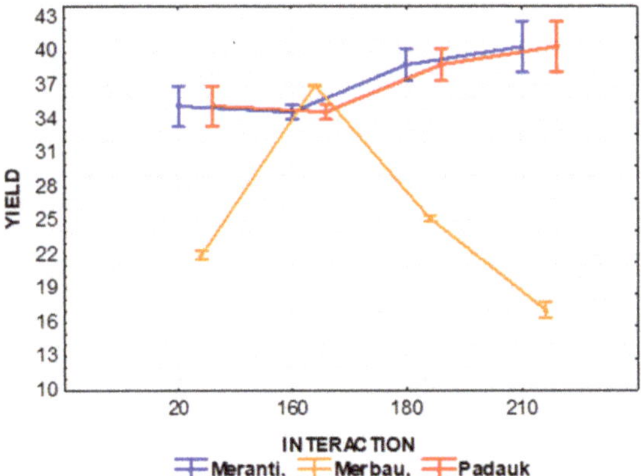

Figure 12. The effect of thermal modification temperature and wood species on YIELD values.

Figure 7 shows the effect of the interaction of the thermal modification temperature and the wood species on M_n values. It is clear from the results that the lowest M_n values were measured in padauk wood, and the highest ones were measured in merbau wood. The effect of temperature on the values of this characteristic was also very significant. In meranti wood, we recorded a significant increase at a temperature of 160 °C, while there was no significant difference between M_n values measured at 20, 180, and 210 °C. In merbau wood, we can see a significant decrease in M_n values with an increase in the temperature of thermal modification.

In padauk wood this trend is not so clear-cut. The effect of a temperature of 160 °C increased the values of the examined characteristic, at a temperature of 180 °C there was a significant decrease in the values, and at 210 °C the values of the monitored characteristic increased again.

Figure 8 shows the interaction of the thermal modification temperature and the wood species on M_w values. The effect of the wood species is similar to the previous case, the lowest M_w values were measured in padauk wood and the significantly highest values in merbau wood. In the case of padauk wood, the values of the monitored characteristic increased significantly with an increase in the applied temperature. In meranti wood, this effect is opposite, and with an increase in the operating temperature, the values of the observed characteristic of M_w decrease significantly. An interesting course can be observed in merbau wood, while the effect of 160 °C cannot be considered significant compared to thermally untreated wood. Temperatures of 180 and 210 °C resulted in a very significant increase in the value of the observed characteristic M_w.

Figure 9 shows the effect of the interaction of the temperature of thermal modification and wood species on M_z values. It is evident from the value in the graph that the lowest values of the observed characteristic were measured in padauk wood again, and the highest values of the monitored characteristic were measured in merbau wood. In padauk wood, a decrease in the values of the observed characteristic M_z can be observed at a temperature of 160 °C in comparison with the untreated wood. With higher temperatures (180 and 210 °C), we achieved an increase in the values of the observed characteristics. In meranti wood, the effect was the opposite. At 160 °C the values of the monitored characteristic increased insignificantly, but at temperatures of 180 and 210 °C there was a very significant decrease in the values of the monitored characteristic. The effect of the interaction of the thermal modification temperature and the wood species on the values of the observed characteristic PDI is very similar as in the previous case (Figure 10).

The interaction of thermal modification temperature and wood species was shown to have an insignificant effect on the monitoring of this interaction on S/G values measured in padauk and meranti wood (Figure 11). While the effect of the monitored interaction had a very significant effect on the values measured in merbau wood, the values measured in this wood were significantly higher than in the case of meranti and padauk wood. We can also see a significant increase in the values of the monitored characteristic at 160 °C, accompanied by a significant decrease in these values at higher temperatures (180 and 210 °C).

YIELD values measured in padauk and meranti wood do not differ. In both woods, an insignificant increase in the values of the monitored characteristic can be observed due to the thermal modification temperature. In contrast, the YIELD values measured in merbau wood decrease significantly under the effect of thermal modification temperature (Figure 12).

Figure 13 shows the relationship between the observed interaction (thermal modification temperature and wood species) and its influence on the monitored characteristics M_n, M_w, M_z, S/G, and YIELD. It is clear from the values of the correlation coefficient and the data in Figure 13 that the effect of the observed interaction has a significant correlation in all monitored cases, and each of the monitored characteristics has a declining trend as a result of the interaction.

Variable	Interaction
M_n	-0.910629
M_w	-0.678771
M_z	-0.540281
PDI	-0.476729
S/G	-0.662953
YIELD	0.649132

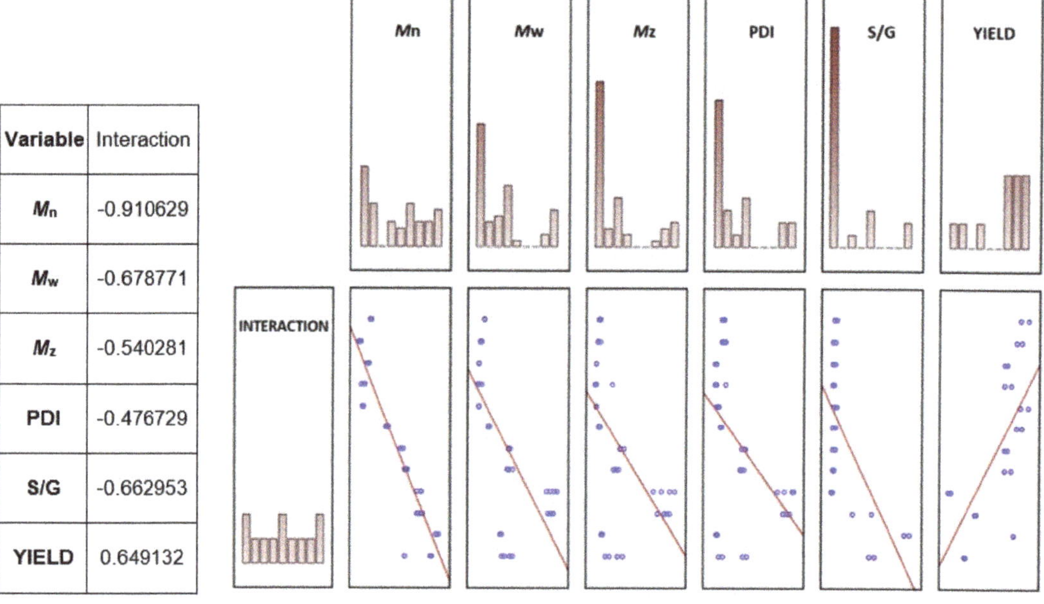

Figure 13. Correlations of the interaction of thermal modification temperature and wood species with the values of the monitored characteristics M_n, M_w, M_z, S/G, and YIELD evaluating the mutual connection between the monitored relationships.

4. Conclusions

Meranti, padauk, and merbau wood samples were thermally treated according to the ThermoWood process at different temperatures (160, 180, 210 °C), lignin was extracted by dioxane and its changes were evaluated by chemical, chromatographic, and spectroscopic analyses. Different changes in all the examined lignins were observed. However, simultaneous degradation and condensation reactions take place during the thermal modification of wood samples. In meranti lignin, the loss of molecular weight occurs, in padauk only the negligible increased, but merbau lignin dramatically condensed at the temperature above 180 °C. All the examined wood species are characterized by an SG type of lignin, with more syringyl units compared to guaiacyl, and with a minor content of hydroxyphenyl-units.

Nitrobenzene oxidation (NBO) analyses show a similar S/G ratio and yield products in meranti and padauk lignins, the higher S/G ratio and products yield at the temperature of 160 °C was found in merbau lignin, then these values rapidly decreased. Infrared spectra also indicated significant changes in lignin structure, which is most pronounced especially in condensation reactions at the highest temperature value. The knowledge about the changes of lignin structure during the heat treatment could be useful in the ThermoWood treatment, recycling of thermally treated wood or in its processing at the end of the life cycle.

Author Contributions: D.K. and F.K. conceived and designed the experiments; M.G. and D.K. prepared samples of isolated wood polymers; I.K. measured and analysed FTIR spectra; F.K. measured and analysed molecular weight distribution; D.K., I.K., and F.K. calculated dependencies from obtained data; D.K., I.K., M.G., and F.K. wrote the paper. All authors have read and agreed to the published version of the manuscript.

Funding: This work was supported by the Slovak Research and Development Agency under contracts no. APVV-17-0005 (40%) and no. APVV-16-0326 (30%), and by the VEGA Agency of Ministry of Education, Science, Research, and Sport of the Slovak Republic no. 1/0387/18 (30%).

Institutional Review Board Statement: Not applicable.

Informed Consent Statement: Not applicable.

Data Availability Statement: Data sharing not applicable.

Conflicts of Interest: The authors declare no conflict of interest. The funders had no role in the design of the study; in the collection, analyses, or interpretation of data; in the writing of the manuscript, and in the decision to publish the results.

References

1. Lee, S.H.; Ashaari, Z.; Lum, W.C.; Halip, J.A.; Ang, A.F.; Tan, L.P.; Chin, K.L.; Tahir, P.M. Thermal treatment of wood using vegetable oils: A review. *Constr. Build. Mater.* **2018**, *181*, 408–419. [CrossRef]
2. Sandak, A.; Sandak, J.; Petrillo, M.; Grossi, P.; Brzezicki, M. Performance of modified wood. In *Wood Modification in Europe: A State–of–Art about Processes, Products and Applications*; Jones, D., Sandberg, D., Goli, G., Todaro, L., Eds.; Firenze University Press: Florence, Italy, 2019; pp. 27–33.
3. Antov, P.; Savov, V.; Krišťák, Ľ.; Réh, R.; Mantanis, G.I. Eco-Friendly, High-Density Fiberboards Bonded with Urea-Formaldehyde and Ammonium Lignosulfonate. *Polymers* **2021**, *13*, 220. [CrossRef]
4. Zhang, P.; Wie, Y.; Liu, Y.; Gao, J.; Chen, Y.; Fan, Y. Heat-Induced Discoloration of Chromophore Structures in Eucalyptus Lignin. *Materials* **2018**, *11*, 1686. [CrossRef] [PubMed]
5. Dos Santos, D.V.B.; De Moura, L.F.; Brito, J.O. Effect of heat treatment on color, weight loss, specific gravity and equilibrium moisture content of two low market valued tropical woods. *Wood Res.* **2014**, *59*, 253–264.
6. Rodriguez-Jimenez, S.; Duarte-Aranda, S.; Canche-Escamilla, G. Chemical composition and thermal properties of tropical wood from the Yucatán dry forests. *BioResources* **2019**, *14*, 2651–2666.
7. Xu, J.; Zhang, Y.; Shen, Y.; Li, C.; Wang, Y.; Ma, Z.; Sun, W. New perspective on wood thermal modification: Relevance between the evolution of chemical structure and physical-mechanical properties, and online analysis of release of VOCs. *Polymers* **2019**, *11*, 1145. [CrossRef] [PubMed]
8. Shi, J.L.; Kocaefe, D.; Amburgey, T.; Zhang, J. A comparative study on brown rot fungus decay and subterranean termite resistance of thermally-modified and ACQ-C-treated wood. *Eur. J. Wood Wood Prod.* **2007**, *65*, 353–358. [CrossRef]
9. Korkut, S. Performance of three thermally treated tropical wood species commonly used in Turkey. *Ind. Crops Prod.* **2012**, *36*, 355–362. [CrossRef]
10. Ghadge, K.; Pandey, K.K. Effect of Thermal Modification on Physical Properties of Bambusa nutans. In *Wood Is Good*; Pandey, K.K., Ramakantha, V., Shakti Chauhan, S.S., Arun Kumar, A.N., Eds.; Springer: Singapore, 2017; pp. 287–295. [CrossRef]
11. Sikora, A.; Kačík, F.; Gaff, M.; Vondrová, V.; Bubeníková, T.; Kubovský, I. Impact of thermal modification on color and chemical changes of spruce and oak wood. *J. Wood Sci.* **2018**, *64*, 406–416. [CrossRef]
12. Hill, C.A.S. Wood modification: Chemical, Thermal and Other Processes. In *Wiley Series in Renewable Resources*; Stevens, C.V., Ed.; John Wiley & Sons: Chichester, UK, 2006; 260p.
13. Carrier, M.; Loppinet-Serani, A.; Denux, D.; Lasnier, J.M.; Ham-Pichavant, F.; Cansell, F.; Aymonier, C. Thermogravimetric analysis as a new method to determine the lignocellulosic composition of biomass. *Biomass Bioenergy* **2011**, *35*, 298–307. [CrossRef]
14. Zinovyev, G.; Sulaeva, I.; Podzimek, S.; Rössner, D.; Kilpelainen, I.; Sumerskii, I.; Rosenau, T.; Potthast, A. Getting Closer to Absolute Molar Masses of Technical Lignins. *ChemSusChem* **2018**, *11*, 3259–3268. [CrossRef] [PubMed]

15. Doherty, W.O.S.; Mousavioun, P.; Fellows, C.M. Value-adding to cellulosic ethanol: Lignin polymers. *Ind. Crops Prod.* **2011**, *33*, 259–276. [CrossRef]
16. Papadopoulos, A.N. Advances in Wood Composites. *Polymers* **2020**, *12*, 48. [CrossRef]
17. Pelaez-Samaniego, M.R.; Yadama, V.; Lowell, E.; Espinoza-Herrera, R. A review of wood thermal pretreatments to improve wood composite properties. *Wood Sci. Technol.* **2013**, *47*, 1285–1319. [CrossRef]
18. Hortobágyi, Á.; Pivarčiová, E.; Koleda, P. Holographic Interferometry for Measuring the Effect of Thermal Modification on Wood Thermal Properties. *Appl. Sci.* **2021**, *11*, 2516. [CrossRef]
19. Ditommaso, G.; Gaff, M.; Kačík, F.; Sikora, A.; Sethy, A.; Corleto, R.; Razaei, F.; Kaplan, L.; Kubš, J.; Das, S.; et al. Interaction of technical and technological factors on qualitative and energy/ecological/economic indicators in the production and processing of thermally modified merbau wood. *J. Clean. Prod.* **2020**, *252*, 119793. [CrossRef]
20. Noh, N.I.F.; Ahmad, Z. Heat treatment on keruing and light red meranti: The effect of heat exposure at different levels of temperature on bending strength properties. *IOP Conf. Ser. Mater. Sci. Eng.* **2017**, *271*, 012060. [CrossRef]
21. Gašparík, M.; Gaff, M.; Kačík, F.; Sikora, A. Color and chemical changes in teak (*Tectona grandis* L. f.) and meranti (*Shorea* spp.) wood after thermal treatment. *BioResources* **2019**, *14*, 2667–2683. [CrossRef]
22. Devashankar, S. FTIR, Powder X-RD and DSC Analysis of African Padauk Wood to Elucidate Possible Applications. In *Macromolecular Symposia*; Awasthi, K., Babu, S.B., Eds.; Wiley-VCH: Weinheim, Germany, 2017; pp. 1–7. [CrossRef]
23. Wang, C.; Qin, Y.; Wang, F.; Wang, Z.; Huan, A. Effect of Iron Oxide on the Protective Photochromism of African Padauk. *Adv. Polym. Technol.* **2019**, 1–8. [CrossRef]
24. Kroupa, M.; Gaff, M.; Karlsson, O.; Myronycheva, O.; Sandberg, D. Effects of thermal modification on bending properties and chemical structure of Iroko and Padauk. In Proceedings of the 9th European Conference on Wood Modification, Burgers' Zoo, Arnhem, The Netherlands, 17–18 September 2018; Jos, C., Thomas, H., Bôke, T., Holger, M., Brigitte, J., Jos, G., Eds.; SHR Wageningen: Wageningen, The Netherlands, 2018; pp. 155–161, ISBN 978-90-829466-1-1.
25. Hu, C.; Jiang, G.; Xiao, M.; Zhou, J.; Yi, Z. Effects of heat treatment on water-soluble extractives and color changes of merbau heartwood. *J. Wood Sci.* **2012**, *58*, 465–469. [CrossRef]
26. Liao, Y.; Wang, J.; Lu, Z.; Gu, J.; Hu, C. Effects of heat treatment on durability of merbau heartwood. *BioResources* **2016**, *11*, 426–438. [CrossRef]
27. Malik, J.; Ozarska, B. Mechanical characteristics of impregnated white Jabon wood (*Anthocephalus cadamba*) using merbau extractives and selected polymerised merbau extractives. *Maderas-Cienc. Tecnol.* **2019**, *21*, 573–586. [CrossRef]
28. Malik, J.; Santoso, A.; Ozarska, B. Polymerised merbau extractives as impregnating material for wood properties enhancement. *IOP Conf. Series: Mater. Sci. Eng.* **2020**, *935*. [CrossRef]
29. Makovická-Osvaldová, L.; Gašparík, M.; Castellanos, J.R.S.; Markert, F.; Kadlicová, P.; Čekovská, H. Effect of thermal treatment on selected fire safety features of tropical wood. *Commun.-Sci. Lett. Univ. Žilina* **2018**, *20*, 3–7.
30. Kamboj, G.; Gašparík, M.; Gaff, M.; Kačík, F.; Sethy, A.K.; Corleto, R.; Razaei, F.; Ditommaso, G.; Sikora, A.; Kaplan, L.; et al. Surface quality and cutting power requirement after edge milling of thermally modified meranti (*Shorea* spp.) wood. *J. Build. Eng.* **2020**, *29*, 101213. [CrossRef]
31. ASTM. *Standard Test Method for Ethanol-Toluene Solubility of Wood*; ASTM D1107-96(2013); ASTM International: West Conshohocken, PA, USA, 2013. [CrossRef]
32. Sluiter, A.; Hames, B.; Ruiz, R.; Scarlata, C.; Sluiter, J.; Templeton, D.; Crocker, D. *Determination of Structural Carbohydrates and Lignin in Biomass*; NREL/TP-510-42618; Laboratory Analytical Procedure (LAP), National Renewable Energy Laboratory: Golden, CO, USA, 2012. Available online: http://www.nrel.gov/biomass/analytical_procedures.html (accessed on 20 February 2021).
33. Kačíková, D.; Kubovský, I.; Ulbriková, N.; Kačík, F. The Impact of Thermal Treatment on Structural Changes of Teak and Iroko Wood Lignins. *Appl. Sci.* **2020**, *10*, 5021. [CrossRef]
34. Rousset, P.; Lapierre, C.; Pollet, B.; Quirino, W.; Perre, P. Effect of severe thermal treatment on spruce and beech wood lignins. *Ann. For. Sci.* **2009**, *66*, 110. [CrossRef]
35. Shinde, S.D.; Meng, X.; Kumar, R.; Ragauskas, A.J. Recent advances in understanding the pseudo-lignin formation in a lignocellulosic biorefinery. *Green Chem* **2018**, *20*, 2192–2205. [CrossRef]
36. Dahali, R.; Lee, S.H.; Ashaari, Z.; Bakar, E.S.; Ariffin, H.; Khoo, P.S.; Bawon, P.; Salleh, Q.N. Durability of Superheated Steam-Treated Light Red Meranti (*Shorea* spp.) and Kedondong (*Canarium* spp.) Wood against White Rot Fungus and Subterranean Termite. *Sustainability* **2020**, *12*, 4431. [CrossRef]
37. Windeisen, E.; Wegener, G. Behaviour of lignin during thermal treatments of wood. *Ind. Crops Prod.* **2008**, *27*, 157–162. [CrossRef]
38. Syafii, W. The effect of lignin composition on delignification rate of some tropical hardwoods. *Indones. J. Trop. Agric.* **2001**, *10*, 9–13. [CrossRef]
39. Evtuguin, D.V.; Neto, C.P.; Silva, A.M.S.; Domingues, P.M.; Amado, F.M.L.; Robert, D.; Faix, O. Comprehensive study on the chemical structure of dioxane lignin from plantation Eucalyptus globulus wood. *J. Agric. Food. Chem.* **2001**, *49*, 4252–4261. [CrossRef] [PubMed]
40. Aguayo, M.G.; Ruiz, J.; Norambuena, M.; Mendonça, R.T. Structural features of dioxane lignin from Eucalyptus globulus and their relationship with the pulp yield of contrasting genotypes. *Maderas-Cienc. Tecnol.* **2015**, *17*, 625–636. [CrossRef]
41. Rana, R.; Langenfeld-Heyser, R.; Finkeldey, R.; Polle, A. FTIR spectroscopy, chemical and histochemical characterisation of wood and lignin of five tropical timber wood species of the family of Dipterocarpaceae. *Wood Sci. Technol.* **2010**, *44*, 225–242. [CrossRef]

42. Kim, J.Y.; Hwang, H.; Oh, S.; Kim, Y.S.; Kim, U.J.; Choi, J.W. Investigation of structural modification and thermal characteristics of lignin after heat treatment. *Int. J. Biol. Macromol.* **2014**, *66*, 57–65. [CrossRef]
43. Lourenço, A.; Neiva, D.M.; Gominho, J.; Curt, M.D.; Fernández, J.; Marques, A.V.; Pereira, H. Biomass production of four *Cynara cardunculus* clones and lignin composition analysis. *Biomass Bioenergy* **2015**, *76*, 86–95. [CrossRef]
44. Cui, C.; Sadeghifar, H.; Sen, S.; Argyropoulos, D.S. Towards thermoplastic lignin polymers; Part II: Thermal & polymer characteristics of kraft lignin & derivatives. *BioResources* **2012**, *8*, 864–886. [CrossRef]
45. Patil, S.V.; Argyropoulos, D.S. Stable Organic Radicals in Lignin: A Review. *ChemSusChem* **2017**, *10*, 3284–3303. [CrossRef] [PubMed]
46. Bubeníková, T.; Luptáková, J.; Kačíková, D.; Kačík, F. Characterization of macromolecular traits of lignin from heat treated spruce wood by size exclusion chromatography. *Acta Fac. Xylologiae* **2018**, *60*, 33–42.
47. Dörrstein, J.; Scholz, R.; Schwarz, D.; Schieder, D.; Sieber, V.; Walther, F.; Zollfrank, C. Effects of high-lignin-loading on thermal, mechanical, and morphological properties of bioplastic composites. *Compos. Struct.* **2018**, *189*, 349–356. [CrossRef]
48. Taghiyari, H.R.; Hosseini, G.; Tarmian, A.; Papadopoulos, A.N. Fluid Flow in Nanosilver-Impregnated Heat-Treated Beech Wood in Different Mediums. *Appl. Sci.* **2020**, *10*, 1919. [CrossRef]
49. Poletto, M.; Zeni, M.; Zattera, A.J. Effects of wood flour addition and coupling agent content on mechanical properties of recycled polystyrene/wood flour composites. *J. Thermoplast. Compos. Mater.* **2012**, *25*, 821–833. [CrossRef]
50. Mattos, B.D.; Lourençon, T.V.; Serrano, L.; Labidi, J.; Gatto, D.A. Chemical modification of fast-growing eucalyptus wood. *Wood Sci. Technol.* **2015**, *2*, 273–288. [CrossRef]
51. Kačík, F.; Kačíková, D.; Bubeníková, T. Spruce wood lignin alteration after infrared heating at different wood moistures. *Cell. Chem. Technol.* **2006**, *40*, 643–648.
52. Esteves, B.; Marques, A.V.; Domingos, I.; Pereira, H. Chemical changes of heat-treated pine and eucalypt wood monitored by FTIR. *Maderas-Cienc. Tecnol.* **2013**, *15*, 245–258. [CrossRef]
53. Košíková, B.; Sláviková, E.; Sasinková, V.; Kačík, F. The use of various yeast strains for removal of pine wood extractive constituents. *Wood Res.* **2006**, *51*, 47–53.
54. Kubovský, I.; Kačíková, D.; Kačík, F. Structural Changes of Oak Wood Main Components Caused by Thermal Modification. *Polymers* **2020**, *12*, 485. [CrossRef]
55. Leclerc, D.F. Fourier Transform Infrared Spectroscopy in the Pulp and Paper Industry. In *Encyclopedia of Analytical Chemistry*; Meyers, R.A., Ed.; John Wiley & Sons Ltd.: Chichester, UK, 2000; pp. 8361–8388.
56. Mvondo, R.R.N.; Meukam, P.; Jeong, J.; Meneses, D.D.S.; Nkeng, E.G. Influence of water content on the mechanical and chemical properties of tropical wood species. *Results Phys.* **2017**, *7*, 2096–2103. [CrossRef]
57. Li, M.Y.; Cheng, S.C.; Li, D.; Wang, S.N.; Huang, A.M.; Sun, S.Q. Structural characterization of steam-heat treated *Tectona grandis* wood analyzed by FT-IR and 2D-IR correlation spectroscopy. *Chin. Chem. Lett.* **2015**, *26*, 221–225. [CrossRef]
58. Čabalová, I.; Kačík, F.; Lagaňa, R.; Výbohová, E.; Bubeníková, T.; Čaňová, I.; Ďurkovič, J. Effect of thermal treatment on the chemical, physical, and mechanical properties of pedunculate oak (*Quercus robur* L.) wood. *BioResources* **2018**, *13*, 157–170. [CrossRef]
59. Windeisen, E.; Strobel, C.; Wegener, G. Chemical changes during the production of thermo-treated beech wood. *Wood Sci. Technol.* **2007**, *41*, 523–536. [CrossRef]
60. Weiland, J.J.; Guyonnet, R. Study of chemical modifications and fungi degradation of thermally modified wood using DRIFT spectroscopy. *Holz. Roh. Werkst.* **2003**, *61*, 216–220. [CrossRef]
61. Bourgois, J.; Guyonnet, R. Characterization and analysis of torrefied wood. *Wood Sci. Technol.* **1988**, *22*, 143–155. [CrossRef]
62. Kačík, F.; Luptáková, J.; Šmíra, P.; Nasswettrová, A.; Kačíková, D.; Vacek, V. Chemical Alterations of Pine Wood Lignin during Heat Sterilization. *BioResources* **2016**, *11*, 3442–3452. [CrossRef]
63. Faix, O. Fourier transform infrared spectroscopy. In *Methods in Lignin Chemistry*; Lin, S.Y., Dence, C.W., Eds.; Springer: Berlin/Heidelberg, Germany, 1992; Chapter 4.1; pp. 83–109.
64. Boeriu, C.G.; Bravo, D.; Gosselink, R.J.A.; van Dam, J.E.G. Characterisation of structure-dependent functional properties of lignin with infrared spectroscopy. *Ind. Crops Prod.* **2004**, *20*, 205–218. [CrossRef]
65. Kubo, S.; Kadla, J.F. Hydrogen bonding in lignin: A Fourier transform infrared model compound study. *Biomacromolecules* **2005**, *6*, 2815–2821. [CrossRef]
66. Watkins, D.; Nuruddin, M.D.; Hosur, M.; Tcherbi-Narteh, A.; Jeelani, S. Extraction and characterization of lignin from different biomass resources. *J. Mater. Res. Technol.* **2015**, *4*, 26–32. [CrossRef]
67. Cheng, S.; Huang, A.; Wang, S.; Zhang, Q. Effect of Different Heat Treatment Temperatures on the Chemical Composition and Structure of Chinese Fir Wood. *BioResources* **2016**, *11*, 4006–4016. [CrossRef]
68. Jakab, E.; Faix, O.; Till, F. Thermal decomposition of milled wood lignins studied by thermogravimetry/mass spectrometry. *J. Anal. Appl. Pyrol.* **1997**, *40–41*, 171–186. [CrossRef]
69. Stark, N.M.; Yelle, D.J.; Agarwal, U.P. Techniques for Characterizing Lignin. In *Lignin in Polymer Composites*; Faruk, O., Sain, M., Eds.; William Andrew Publishing: Oxford, UK, 2016; pp. 49–66. ISBN 978-0-323-35565-0.

MDPI

Article

Chemical and Morphological Composition of Norway Spruce Wood (*Picea abies*, L.) in the Dependence of Its Storage

Iveta Čabalová [1,*], Michal Bélik [1], Viera Kučerová [1] and Tereza Jurczyková [2]

[1] Department of Chemistry and Chemical Technologies, Faculty of Wood Sciences and Technology, Technical University in Zvolen, T. G. Masaryka 24, 960 53 Zvolen, Slovakia; xbelik@is.tuzvo.sk (M.B.); viera.kucerova@tuzvo.sk (V.K.)

[2] Department of Wood Processing, Czech University of Life Sciences in Prague, Kamýcká 1176, 16521 Praha, Czech Republic; jurczykova@fld.czu.cz

* Correspondence: cabalova@tuzvo.sk; Tel.: +42-1455206375

Abstract: Chemical composition and morphological properties of Norway spruce wood and bark were evaluated. The extractives, cellulose, hemicelluloses, and lignin contents were determined by wet chemistry methods. The dimensional characteristics of the fibers (length and width) were measured by Fiber Tester. The results of the chemical analysis of wood and bark show the differences between the trunk and top part, as well as in the different heights of the trunk and in the cross section of the trunk. The biggest changes were noticed between bark trunk and bark top. The bark top contains 10% more of extractives and 9.5% less of lignin. Fiber length and width depends on the part of the tree, while the average of these properties are larger depending on height. Both wood and bark from the trunk contains a higher content of fines (fibers <0.3 mm) and less content of longer fibers (>0.5 mm) compared to the top. During storage, it reached a decrease of extractives mainly in bark. Wood from the trunk retained very good durability in terms of chemical composition during the storage. In view of the morphological characteristics, it occurred to decrease both average fibers length and width in wood and bark.

Keywords: spruce wood; cellulose; hemicelluloses; lignin; extractives; time of storage; fiber characteristics

Citation: Čabalová, I.; Bélik, M.; Kučerová, V.; Jurczyková, T. Chemical and Morphological Composition of Norway Spruce Wood (*Picea abies*, L.) in the Dependence of Its Storage. *Polymers* **2021**, *13*, 1619. https://doi.org/10.3390/polym13101619

Academic Editor: Salim Hiziroglu

Received: 4 May 2021
Accepted: 13 May 2021
Published: 17 May 2021

Publisher's Note: MDPI stays neutral with regard to jurisdictional claims in published maps and institutional affiliations.

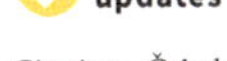

1. Introduction

Wood is an anisotropic material, with respect to its anatomical, physical, and chemical properties, and is made up of different kinds of cells. Wood is degradable by fungi, microorganisms and heating [1]. Degradation of wood and its chemical structure are influenced by the temperature, oxygen available to the material, ambient pressure, wood type and shape, moisture content of the wood, and additives, such as inorganic substances, sorbed emissions, etc. [2–4]. Fedyukov et al. [5] progressively describe the decreasing of cellulose content of spruce wood with soil condition deterioration.

The structure of wood, chemical components (cellulose, hemicelulloses, lignin and extractives) and their relative mass proportion depending on the morphological region, kind of the tree, and age of the wood [1]. Variation can be found within a single tree from the center of the trunk to the bark, from the trunk to the top, between earlywood and latewood, and between sapwood and heartwood [6]. According to Fengel and Wegener [7], earlywood contains more lignin and inversely less cellulose than latewood.

Understanding the morphological and chemical heterogeneity of wood is important in its utilization, for example in the paper industry. Like wood, bark is an important source of raw material and chemical compounds [8]. The chemical composition of bark varies among the different tree species and depends on the morphological element involved. Many of the constituents present in wood also occur in bark, but their proportion is different. Bark can roughly be divided into the fraction: fibers, corok cells and fine substance (including parenchyma cells) [1]. Bark contains much more extractives than wood from the trunk [9], considerable amounts of bioactive components such as antioxidants (polyphenols), and also structural polysaccharides such as pectins [10–12]. Degradation of biomass is influenced by photodegradation (UV light), high temperatures during storage in piles, microorganizms, etc., Routa et al. [13] describe the major reactions of lipophilic extractive compounds during wood storage, which can be divided into three types: (1) hydrolysis of triglycerides (rapid reaction) and steryl esters and waxes (proceeds slower); (2) oxidation/degradation/polymerization of resin acids, unsaturated fatty acids, and to some extent, other unsaturated compounds; and (3) evaporation of volatile terpenoids, mainly monoterpenes.

As any wood material, spruce wood is chemically complex and its physical, chemical and morphological characteristics are not uniform. In Slovakia, it is the most common wood specie. In Norway spruce, approximately 95% of wood cell matrix is composed of tracheids, which can also be termed fibers. The average tracheid length is mainly influenced by tree age. As most wood properties, fibers length and width varies greatly both within and among trees, depending on its vertical and radial position (ring age) in trunk, and forest stand [14]. The tracheids length is shortest next to the pith, the increase with age is at first very rapid in the juvenile period, then slow down between the ages 10–30 and thereafter, as mature wood begins to form, increases very gradually with seasonal fluctuation [15].

The aim of this paper is to characterize chemical composition (extractives, lignin, polysaccharides, cellulose and hemicelluloses) and morphological properties (fibers length and width) of different parts of Norway spruce wood (*Picea abies* (L.) Karst.) and to evaluate the rate of changes in these properties after wood storage.

2. Experimental

2.1. Materials

Norway spruce wood (*Picea abies* L. Karst.) was harvested in the middle part of Slovakia (Zvolen region) in June. The 65-year-old tree was cut 0.5 m from the ground and its height was 25 m. For the experiment, we used a 5 m long trunk part (the diameter of 32 cm—thicker part and the diameter of 29.5 cm—thinner part) and the top part of tree, represented the waste biomass (the diameter of 12 cm—thicker part and the diameter of 7.5 cm—thinner part). We took samples (cut with an all diameter) from the trunk in a height of 0.5, 1.5, 2.5, 3.5 and 4.5 m from the ground and both wood and bark from the top. Equally, we took samples A, B, C (Figure 1) in a height of 0.5 and 4.5 m, whereas sample A represent juvenile wood, which is within the first 20 years of growth; B sample, the next 20 years; and C sample, the rest.

Figure 1. Samples A, B, and C used for analysis.

After trunk sampling, we obtained five wood pieces with a length of approximately 1 m. We debarked the first one. A part of the bark was used for the experiment and the rest for storage. The rest (four) of the wood pieces (in the bark) were also used for storage. After 2, 4, 6 and 8 months of storage, we always analysed one piece of wood (the first sample was wood and the next sample bark from this piece) and bark, which was stored separately. Specifications of the samples are described in Table 1.

Table 1. Samples signification.

wood 0.5 m; 1.5 m; 2.5 m; 3.5 m; 4.5 m	sample taken from the tree trunk in a height of 0.5 m; 1.5 m; 2.5 m; 3.5 m; 4.5 m from the ground
wood-top	sample taken from the tree top
bark-trunk	sample taken from the trunk
bark-top	sample taken from the top
wood 0.5 m-A; 4.5 m-A	sample taken from the tree trunk in a height of 0.5 m or 4.5 m from the ground, first 20-years of growth (Figure 1)
wood 0.5 m-B; 4.5 m-B	sample taken from the tree trunk in a height of 0.5 m or 4.5 m from the ground, from 20 to 40 years of growth (Figure 1)
wood 0.5 m-C; 4.5 m-C	sample taken from the tree trunk in a height of 0.5 m or 4.5 m from the ground, the last 25-years of growth (Figure 1)
wood 0, 2, 4, 6, 8 month	wood from the trunk taken in time of 0, 2, 4, 6, and 8 month of its storage
bark trunk 0, 2, 4, 6, 8 month	bark from the trunk taken in time of 0, 2, 4, 6, and 8 month of its storage
bark 0, 2, 4, 6, 8 month	separately stored bark taken in time of 0, 2, 4, 6, and 8 month of its storage

During storage, the conditions were as follows (Table 2):

Table 2. Conditions during wood storage.

Time of Storage/ Conditions	Average Air Humidity (%)	Average Air Temperature (°C)	Average Precipitation (mm)
from Jun to August	82.99 (max. 99.05, min. 59.32)	19.31 (max. 23.08, min. 14.37)	3.20 (max. 29.0, min. 0)
from August to October	86.89 (max. 99.17, min. 71.39)	17.02 (max. 23.63, min. 8.95)	2.28 (max. 23.80, min. 0)
from October to December	96.98 (max. 99.17, min. 81.52)	5.39 (max. 12.45, min. −2.99)	2.06 (max. 42.2, min. 0)
from December to February	97.45 (max. 99.17, min. 76.56)	0.54 (max. 6.17, min. −8.79)	1.50 (max. 17.8, min. 0)

2.2. Methods

2.2.1. Chemical Composition of Wood

The samples were disintegrated into sawdust, and fractions 0.5 mm to 1.0 mm in size were used for the chemical analyses. The extractives content (EL) was determined in a Soxhlet apparatus with a mixture of ethanol and toluene (2:1) according to ASTM D1107-96 [16]. The lignin content (LIG) was determined according to Sluiter et al. [17], and the cellulose content (CEL) was determined according to the method by Seifert [18], and the holocellulose content according to the method by Wise et al. [19]. Hemicelluloses (HEMI) were calculated as the difference between the holocellulose and cellulose contents. Measurements were performed on four replicates per sample. The results were presented as oven-dry wood percentages.

2.2.2. Fibers Length and Width

Two hundred-millilitre mixtures of concentrated CH_3COOH and 30% H_2O_2 (1:1, *v/v*) were poured onto the wood samples (weight = 10 g and dimensions = 20 mm × 2 mm × 2 mm). Then, the samples were refluxed for 3 h, suction filtered through a sintered glass filter (S1), and washed with distilled water. An L & W Fiber Tester (Lorenzen and Wettre, Kista,

Sweden) was used to determine the fibre dimensional characteristics. This measurement is based on the principle of two-dimensional imaging technology. The measurement technology is automated, allowing for frequent and rapid analysis of the fibre quality. The instrument measures various fibre properties, such as the length and width of the fibres, fine portion (from 0.1 mm to 0.2 mm). Measurements were performed on a single replicate per sample, and the number of fibres within each population of the replicate ranged from 19,182 cells to 21,128 cells.

3. Results and Discussion

3.1. The Differences in the Spruce Wood and Bark in the View of Its Chemical Composition and a both Fibers Length and Width Distribution

Wood is composed of cellulose, hemicelluloses, lignin, and extractives. From the results of the chemical analysis of spruce wood (Figures 2 and 3), we obtained the differences between the wood, as part of the trunk, and the top part of the tree. We compared the chemical composition of the trunk part in a height of 0.5 m and 4.5 m. We have taken samples (A, B, C) from the tree trunk, as shown Figure 1. The results indicate that there are differences in extractives content between A, B, and C in a height of 0.5 m. The biggest content of extractives, 1.68%, is in the part C (the last 25 years of growth), then A, 1.29%, and the last B, 1.04%. In a height of 4.5 m, there are no differences between the A and B part, and the amount of extractives in part C is comparable to the results in a height of 0.5 m. The amount of LIG, CEL and HEMI was very similar in a height of 0.5 and 4.5 m as well (Figure 2).

The analysis of chemical composition of wood and bark in different heights (Figure 3) exposed that the extractive content of both wood and bark is growing up with a height of the tree. The biggest extractives content of 32.35% was recorded in bark taken from the top part. Several authors mention the extractives amount of spruce bark from 23.5% to 28.3%, depending on the part of bark (inner bark from 17.3 to 38.7%, outer from 19.1 to 43.3%) [20–23]. The extractives content of the wood part is between 1% and 4.5% [24,25], while there is a difference between sapwood, values from 1.7% to 2.7% and heartwood, from 1.1% to 1.8% [26].

The amount of main chemical compounds: LIG (from 23.69 to 26.06%), CEL (from 39.01 to 42.51%), and HEMI (from 34.98 to 35.30%) were very similar in wood samples taken from the trunk (in a height from 0.5 to 4.5 m) compared to the top part. Sjostrom [27] mentions the lignin content of 27.5%, cellulose of 39.5 and hemicelluloses of 17.2% for Picea glauca; Wang et al. [28] showed a Klasson lignin of 28.2% for *Picea abies* (L.) H. Karst; Neiva et al. [29] showed a lignin content of 27.22%, polysaccharides of 71.18% for *Picea abies* (L.) H. Karst; Harris [30] showed a lignin from 28.0% to 30.8%, cellulose content from 38.1% to 40.3% for juvenile wood, and the lignin from 26.1% to 28.2%, cellulose from 40.2% to 42.7% for the mature wood of Norway spruce.

The biggest changes in chemical composition we noticed between bark trunk and bark top (Figure 3). As mentioned above, the extractive content was higher in the bark-top (32.35%) compared to the bark-trunk (22.16%), the lignin content was higher in bark-trunk (24.97%) compared to bark-top (15.46%), and the content of polysaccharides were similar (52.87% bark-trunk and 52.18% bark-top). Neiva et al. [29] determined the lignin between 26.86% and 29.92%, and polysaccharides content between 37.86% and 52.27%, depending on the bark fraction.

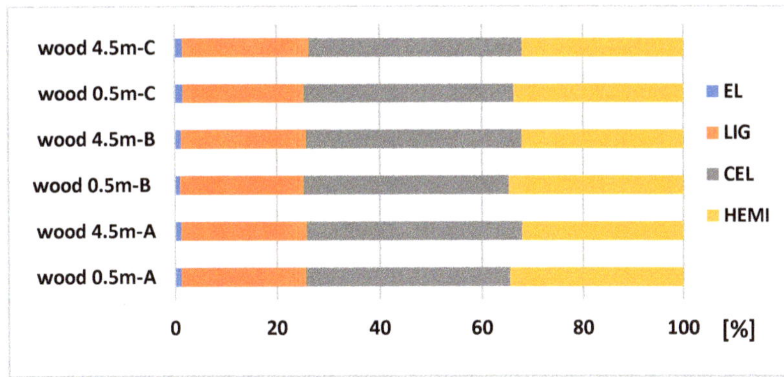

Figure 2. Chemical composition (relative values) of spruce wood trunk at heights of 0.5 and 4.5 m.

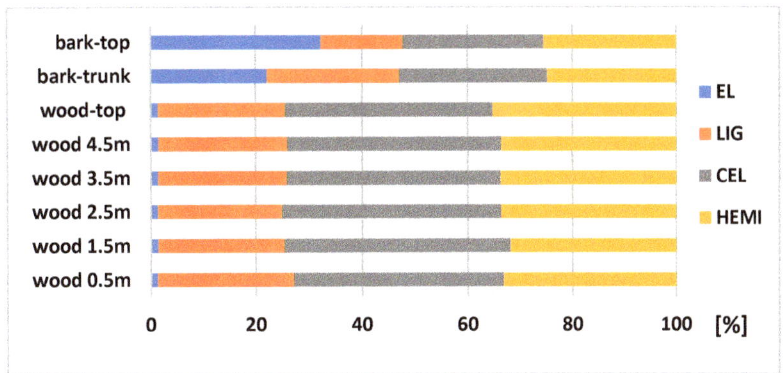

Figure 3. Chemical composition (relative values) of spruce wood and bark at different heights.

Results of the Fiber Tester analysis (Figures 4 and 5) show that there are differences in both wood fibres length and width, mainly between samples A and B in a height of 0.5 and 4.5 m. The lower part of the trunk contains the highest amount of fines (>0.3 mm), from 45.94% (part B) to 52.38% (part A). The lowest content of long fibers (more than 1.01 mm) we determined in sample 0.5-A. There are also visible differences of fibers width in this sample (Figure 5). This part of the wood contains a higher amount of narrow fibers compared to sample 4.5-A. The average fiber length depends on the part of the trunk. The results show (Table 3) the smallest average fiber length and width in sample A, then B, and lastly C in a height of 0.5 m and 4.5 m as well. Both average fiber length and width are larger with the height of the tree.

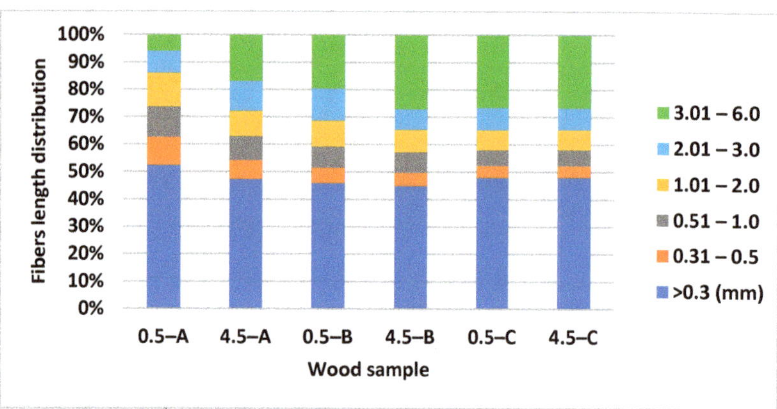

Figure 4. Fibers length distribution of spruce wood trunk at heights of 0.5 and 4.5 m. Fibres length classes: >0.3; 0.31–0.5; 0.51–1.0; 1.01–2.0; 2.01–3.0; 3.01–6 mm.

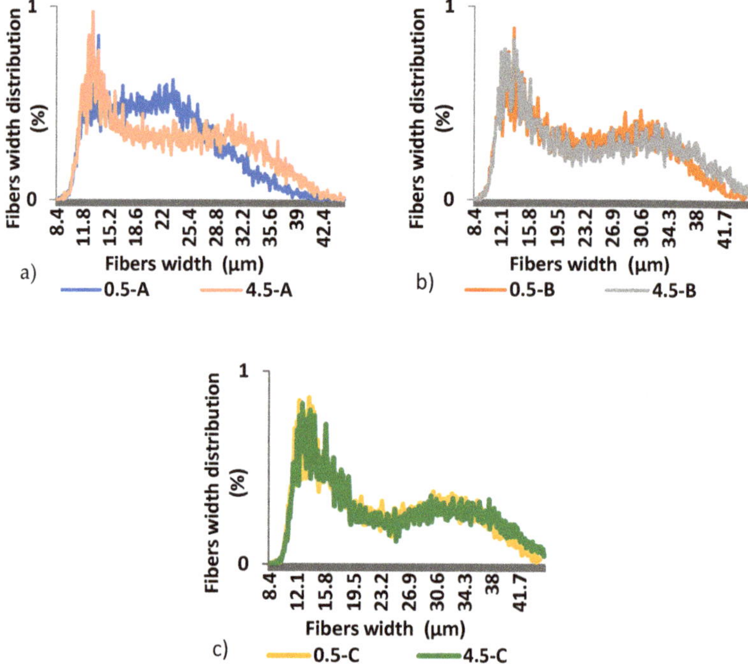

Figure 5. Fibers width distribution of spruce wood trunk at heights of 0.5 and 4.5 m; (**a**) samples 0.5-A, 4.5-A; (**b**) samples 0.5-B, 4.5-B; (**c**) samples 0.5-C, 4.5-C.

Wood from the trunk contains the highest amount of fines, approximately 50% (Figure 4) and its average length is 1.5 mm (Table 4). Tyrväinen [31] presents the average fiber length and its variation of Norway spruce trunk, namely inner heartwood 1.9 mm (1.28–2.70 mm), middle zone 3.0 mm (1.69–3.88 mm), and outer sapwood 3.7 mm (2.80–4.29 mm). Harris [30] and Lönnberg et al. [32] mentioned the fiber width of Norway spruce as being between 15.0 and 28.5 μm in juvenile wood, and between 29.3 and 39.7 μm in mature wood. According to our results (Figure 6), the biggest content of fraction (from

45% wood-trunk to 60.85% wood-top) is in the width class from 15.1 to 30.0 μm and the average width of the wood trunk is 24.33 μm and bark 21.02 μm (Table 4).

The results of the bark analysis (Figures 6 and 7) show the differences mainly in fibre length distribution. The results of fibers width distribution were very similar and the largest proportion of fibers is in the class from 15.1 to 30 μm (67.53–69.17%). Bark from the trunk contains a high amount of fines (65.9%) and a lower amount of longer fibers compared to the bark from the top part of the tree. Bark contains a higher amount of shorter fibers to 0.5 mm (65.88% bark-top, 80.81% bark-trunk), than the wood part.

Table 3. The average fibre length and width of wood samples.

Trait/Sample	0.5-A	0.5-B	0.5-C	4.5-A	4.5-B	4.5-C
average fiber length (mm)	0.83	1.33	1.55	1.22	1.61	1.65
average fiber width (μm)	21.58	23.58	23.91	23.26	24.49	24.70

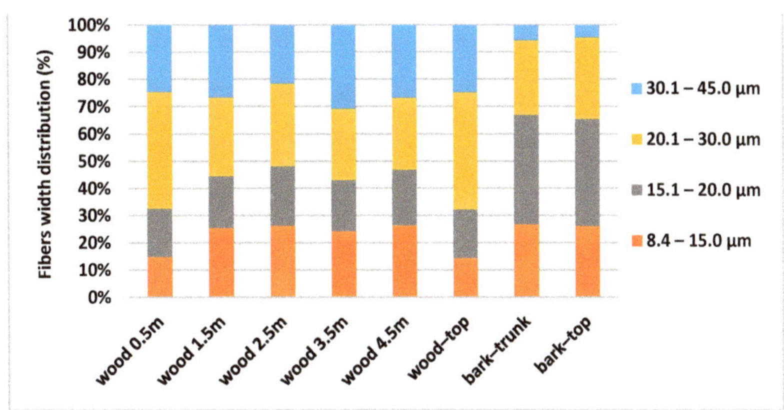

Figure 6. Fibres width distribution of spruce wood and bark at different heights. Fibres width classes: 8.4–15.0 μm; 15.1–20.0 μm; 20.1–30.0 μm; 30.1–45.0 μm.

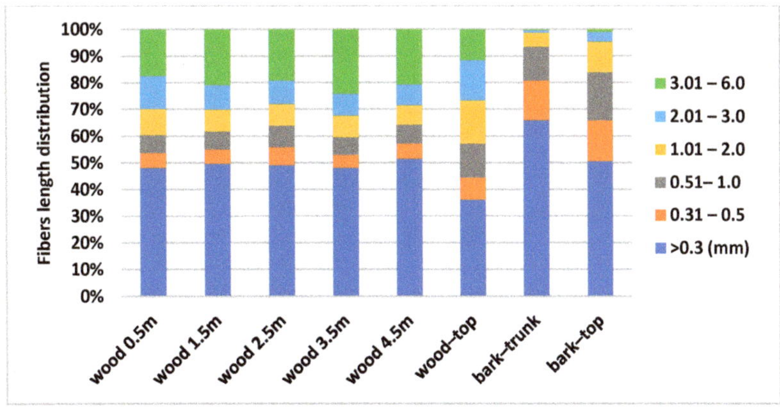

Figure 7. Fibres length distribution of spruce wood and bark at different heights.

3.2. The Differences in the Spruce Wood and Bark after Storage

The amount of extractives is influenced by wood storage. Tree bark is a biological, heterogeneous material whose composition is changing. Immediately after tree harvesting, the amount of volatiles declined, and degradation continued during wood storage.

According to the results (Figures 8 and 9), it is visible that the biggest changes in the chemical composition can be obtained in bark (stored separately and as a part of the trunk). Figure 10 explains the results of chemical composition of wood trunk, while values are very comparable. Spruce wood retained very good durability in terms of its chemical composition after eight months of storage. Several authors studied wood with different natural aging time and the results found that the proportion of saccharides gradually decreases (mainly due to the hemicelluloses degradation), and the content of lignin increases successively with increasing time [33–35].

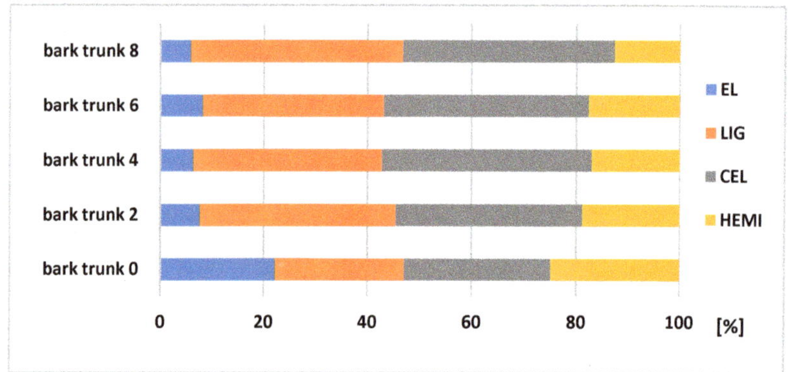

Figure 8. Changes in the composition (relative values) of spruce bark from trunk after its storage of 2, 4, 6, and 8 months.

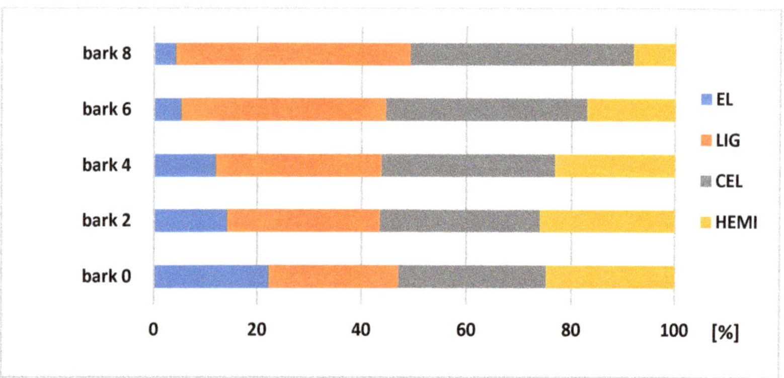

Figure 9. Changes in the composition (relative values) of spruce bark after its storage of 2, 4, 6, and 8 months.

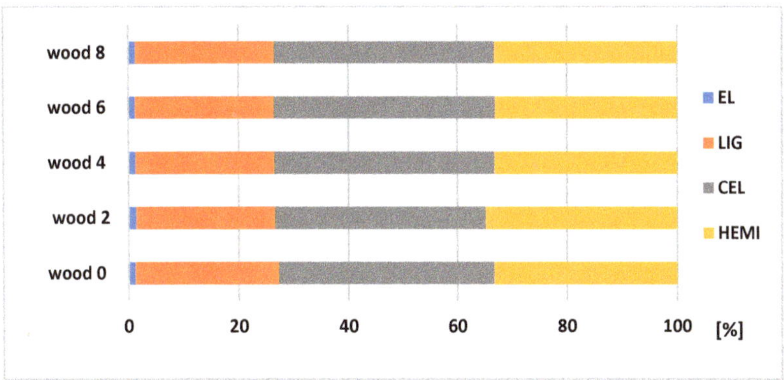

Figure 10. Changes in the spruce wood composition (relative values) after its storage of 2, 4, 6, and 8 months.

The degradation of main chemical components of both bark stored separately and bark as a part of the trunk, was obvious. During eight months of storage, a decreasing of extractives occurred (bark of 80.96%, bark trunk of 73.69%) and HEMI (bark of 67.52%, bark trunk of 49%), and increasing of LIG (bark of 45.65%, bark trunk of 60.9%) and CEL (bark of 65.73% and bark trunk of 69.11%). Bergström and Matison [36] in their study describe a decrease in the extractive content during storage, roughly halving during the first four weeks. The biggest losses noticed during this period were in the amounts of hydrophilic, and lipophilic as well. From chemical components, the stilbenes are very sensitive to degradation [37]. According to Routa et al. [13], after eight weeks of pine bark storage, we obseved types of extractive substances, which are predominate in the bark: triglycerides, steryl esters, sterols, resin acids, and fatty acids.

The Fiber Tester analysis (Figures 11 and 12) shows changes in fibers length and width distribution of both wood and bark. Sample of wood stored for eight month contains lower amount of fibers longer than 1.01 mm (decrease of 23,9%). The both average fiber length and width of samples decreased during storage (Table 4) because of wood and bark degradation.

Table 4. The average fibre length and width before and after storage for 8 months.

Trait/Sample	Wood-Trunk	Wood-Trunk 8	Bark-Trunk	Bark 8	Bark-Trunk 8
average fiber length (mm)	1.50	1.02	0.49	0.35	0.38
average fiber width (µm)	24.33	22.50	21.02	19.57	19.36

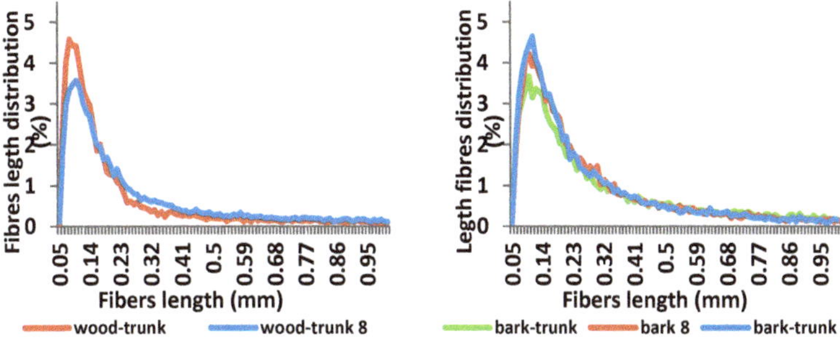

Figure 11. Fibres length distribution of spruce wood trunk and bark before and after its storage of 8 months.

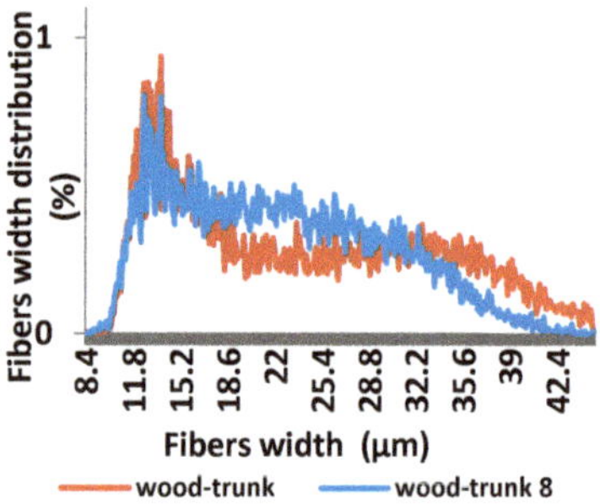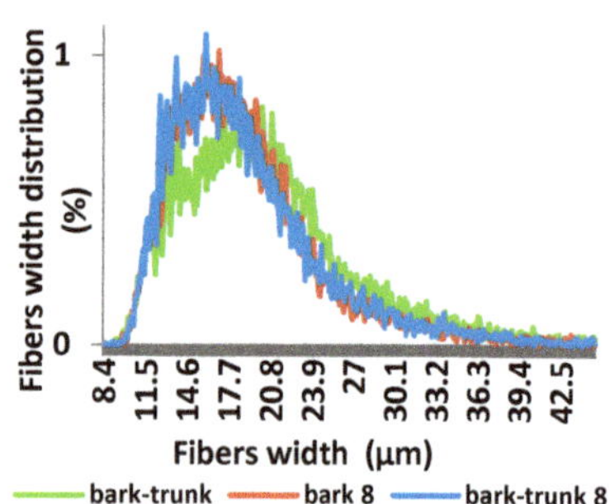

Figure 12. Fibres width distribution of spruce wood trunk and bark before and after storage for 8 months.

4. Conclusions

1. The amount of extractives is the highest in the last 25 years of growth (close to bark), polysaccharides and lignin content was similar in a cross section of trunk.
2. Compared to the chemical composition of wood trunk in a different height (0.5, 1.5, 2.5, 3.5 and 4.5 m from the ground), very similar results of cellulose, hemicelluloses and lignin content were determined. The extractive content of both wood and bark is growing up with the height of the tree.
3. From the view of chemical composition, the differences between bark, as a part of the trunk and the top part of the tree were obtained. In the bark top we determined bigger amount of extractives (10%) and less amount of lignin (9.5%) compared to the bark trunk. The amount of polysaccharides was similar.
4. Juvenile wood contains a smaller amount of longer fibers, >1.01 mm, than other parts of the tree.
5. The both average fiber length and width is higher with a height of the tree.
6. The amount of extractives is very influenced by the time of storage, especially in bark. During eight months of storage, a decreasing of extractives occurred (from 73.7 to 80.9%) and hemicelluloses (from 49.1 to 67.5%) relative content, and an increasing of lignin and cellulose. Bark stored separately degraded faster than bark stored on the trunk.
7. In view of the chemical composition of the wood from the trunk retained very good durability during the storage for eight months.
8. Eight months stored wood contains a lower amount of fibers, longer than 1.01 mm compared to raw wood. The both average fibers length and width decreased during storage.

Author Contributions: I.Č. conceived and designed the experiments; M.B., V.K. and T.J. carried out the laboratory experiments; I.Č. analyzed the data, interpreted the results, prepared figures, and wrote the manuscript. All authors have read and agreed to the published version of the manuscript.

Funding: This research received no external funding.

Acknowledgments: This work was supported by funding from Slovak Scientific Grant Agency VEGA (Contract 1/0397/20 (100%)).

Conflicts of Interest: The authors declare no conflict of interest.

References

1. Yang, G.; Jaakkola, P. *Wood Chemistry and Isolation of Extractives from Wood. Literature Study for BIOTULI Project*; Saimaa University of Applied Sciences: South Karelia, Finland, 2011; 47p.
2. Militz, H. *Thermal Treatment of Wood: European Processes and Their Background (IRG/WP 02-40241)*; International Research Group on Wood Preservation: Stockholm, Sweden, 2002.
3. Hill, C.A.S. *Wood Modification-Chemical, Thermal and Other Processes*; John Wiley & Sons Ltd.: Chicester, UK, 2006.
4. Reinprecht, L. *Wood Deterioration, Protection and Maintenance*; John Wiley & Sons: Chichester, UK, 2016.

5. Fedyukov, V.I.; Boyarsky, M.V.; Saldaeva, E.Y.; Chernova, M.S.; Chernov, V.Y. Dependence of spruce wood resonance properties on its chemical composition. *Wood Res.* **2018**, *63*, 887–894.

6. Côté, W.A. Chemical Composition of Wood. In *Principles of Wood Science and Technology*; Springer: Berlin/Heidelberg, Germany, 1968.

7. Fengel, D.; Wegener, G. *Wood-Chemistry, Ultrastructure, Reactions*, 2nd ed.; Walter de Gruyter: Berlin, Germany, 1989; 613p.

8. Réh, R.; Krišťák, L.; Sedliačik, J.; Bekhta, P.; Božiková, M.; Kunecová, D.; Vozárová, V.; Tudor, E.M.; Antov, P.; Savov, V. Utilization of birch bark as an eco-friendly filler in urea-formaldehyde adhesives for plywood manufacturing. *Polymers* **2021**, *13*, 511. [CrossRef] [PubMed]

9. Rhén, C. Chemical composition and gross calorific value of the above-ground biomass components of young Picea abies. *Scan. J. For. Res.* **2004**, *19*, 72–81. [CrossRef]

10. le Normand, M.; Rietzler, B.; Vilaplana, F.; Ek, M. Macromolecular Model of the Pectic Polysaccharides Isolated from the Bark of Norway Spruce (*Picea abies*). *Polymers* **2021**, *13*, 1106. [CrossRef]

11. Kemppainen, K.; Siika-Aho, M.; Pattathil, S.; Giovando, S.; Kruus, K. Spruce bark as an industrial source of condensed tannins and non-cellulosic sugars. *Ind. Crops Prod.* **2014**, *52*, 158–168. [CrossRef]

12. Jablonský, M.; Vernarecová, M.; Ház, A.; Dubinyová, L.; Škulcová, A.; Sladková, A.; Šurina, I. Extraction of phenolic and lipophilic compounds from spruce (*Picea abies*) bark using accelerated solvent extraction by ethanol. *Wood Res.* **2015**, *6*, 583–590.

13. Routa, J.; Brännström, H.; Hellström, J.; Laitila, J. Influence of storage on the physical and chemical properties of Scots pine bark. *Bioenergy Res.* **2021**, *14*, 575–587. [CrossRef]

14. Dinwoodie, J.M. Tracheid and fibre length in timber: A review of literature. *Forestry* **1961**, *34*, 125–144. [CrossRef]

15. Kucera, B. Juvenile wood formation in Norway spruce. *Wood Fiber Sci.* **1994**, *26*, 152–167.

16. ASTM D1107-96. *Standard Test Method for Ethanol-Toluene Solubility of Wood*; ASTM International: West Conshohocken, PA, USA, 2007.

17. Sluiter, A.; Hames, B.; Ruiz, R.; Scarlata, C.; Sluiter, J.; Templeton, D.; Crocker, D. *Determination of Structural Carbohydrates and Lignin in Biomass*; (NREL/TP-510-42618); National Renewable Energy Laboratory: Golden, CO, USA, 2012.

18. Seifert, V.K. Über ein neues Verfahren zur schnell Bestimmung der rein-Cellulose [About a new method for rapid determination of pure cellulose]. *Das Pap.* **1956**, *10*, 301–306.

19. Wise, L.E.; Murphy, M.; D'Addieco, A.A. Chlorite holocellulose, its fractionation and bearing on summative wood analysis and on studies on the hemicelluloses. *Pap. Trade J.* **1946**, *122*, 35–44.

20. Ånäs, E.; Ekman, R.; Holmbom, B. Composition of nopolar extractives in bark of Norway spruce and Scot pine. *J. Wood Chem. Technol.* **1983**, *3*, 119–130. [CrossRef]

21. Tyalor, A.M.; Gartner, B.L.; Morrell, J.J. Heartwood formation and natural durability—A review. *Wood Fiber Sci.* **2002**, *34*, 587–611.

22. Wajs, A.; Pranovich, A.; Reunanen, M.; Willför, S.; Holmbom, B. Characterisation of volatile organic compounds in stemwood using solid-phase microextraction. *Phytochem. Anal. Int. J. Plant Chem. Biochem. Tech.* **2006**, *17*, 91–101. [CrossRef] [PubMed]

23. Barnett, J.; Jeronimidis, G. *Wood Quality and Its Biological Basis*; John Wiley &Sons: Hoboken, NJ, USA, 2009; 240p.

24. Anttonen, S.; Piispanen, R.; Ovaska, J.; Mutikainen, P.; Saranpää, P.; Vapaavuori, E. Effects of defoliation on growth, biomass allocation, and wood properties of betula clones grown at different nutrient levels. *Can. J. For. Res.* **2002**, *32*, 498–508. [CrossRef]

25. Kaakinen, S.; Piispanen, R.; Lehto, S.; Metsometsä, J.; Nilson, U.; Saranpää, P.; Linder, S.; Vapaavuori, E. Growth, wood chemistry, and fibre length of Norway spruce in long-term nutrient optimization experiment. *Can. J. For. Res.* **2009**, *39*, 410–419. [CrossRef]

26. Hakkila, P. *Utilization of Residual Forest Biomass*; Springer: Berlin/Heidelberg, Germany, 2011; 568p.

27. Sjostrom, E. *Wood Chemistry. Fundamentals and Applications*, 2nd ed.; Academic Press: San Diego, CA, USA, 1993; 292p.

28. Wang, Z.; Winestrand, S.; Gillgren, T.; Jönsson, L.J. Chemical and structural factors influencing enzymatic saccharification of wood from aspen, birch and spruce. *Biomass Bioenergy* **2018**, *109*, 125–134. [CrossRef]

29. Neiva, D.M.; Araújo, S.; Gominho, J.; Carneiro, A.D.C.; Pereira, H. An integrated characterization of *Picea abies* industrial bark regarding chemical composition, thermal properties and polar extracts activity. *PLoS ONE* **2018**, *13*, e0208270. [CrossRef]

30. Harris, G. Comparison of northern softwood and southern pine fiber characteristics for groudwood publication paper. *TAPPI* **1993**, *76*, 55–61.

31. Tyrväinen, J. Wood and Fiber Properties of Norway spruce and its suitability for thermomechanical pulping. *Acta For. Fenn.* **1995**, *249*, 210–263. [CrossRef]

32. Lönberg, B.; Bruun, H.; Lindquist, J. UV-microspectrophotometric study of wood and fibers. *Pap. Wood* **1991**, *73*, 848–851.

33. Inari, G.N.; Petrissans, M.; Lambert, J.; Ehrhardt, J.J.; Gérardin, P. XPS characterization of wood chemical composition after heat-treatment. *Surf. Interface Anal.* **2006**, *38*, 1336–1342. [CrossRef]

34. Popescu, C.-M.; Hill, C.A.S. The water vapour adsorption-desorption behaviour of naturally aged Tilia cordata Mill. wood. *Polym. Degrad. Stab.* **2013**, *98*, 1804–1813. [CrossRef]

35. Zhao, C.; Zhang, X.; Liu, L.; Yu, Y.; Zheng, W.; Song, P. Probing Chemical Changes in Holocellulose and Lignin of Timbers in Ancient Buildings. *Polymers* **2019**, *11*, 809. [CrossRef]

36. Bergström, D.; Matison, M. *Efficient Forest Biomass Supply Chain Management for Biorefineries. Synthesis Report. Forest Refine 2012–2014*; SP Processum: Hörnberg, Andreas; Joelsson, Jonas; Sveriges lantbruksuniversitet: Umeå, Sweden, 2014; 116p.

37. Holmbom, B. Extraction and utilisation of non-structural wood and bark components. *Biorefining For. Resour.* **2011**, *20*, 178–224.

Article

Eco-Friendly, High-Density Fiberboards Bonded with Urea-Formaldehyde and Ammonium Lignosulfonate

Petar Antov [1,*], Viktor Savov [1], Ľuboš Krišťák [2,*], Roman Réh [2] and George I. Mantanis [3]

1   Department of Mechanical Wood Technology, Faculty of Forest Industry, University of Forestry, 1797 Sofia, Bulgaria; victor_savov@ltu.bg
2   Faculty of Wood Sciences and Technology, Technical University in Zvolen, T. G. Masaryka 24, 960 01 Zvolen, Slovakia; reh@tuzvo.sk
3   Department of Forestry, Wood Sciences and Design, Lab of Wood Science and Technology, University of Thessaly, 43100 Karditsa, Greece; mantanis@uth.gr
*   Correspondence: p.antov@ltu.bg (P.A.); kristak@tuzvo.sk (Ľ.K.)

Abstract: The potential of producing eco-friendly, formaldehyde-free, high-density fiberboard (HDF) panels from hardwood fibers bonded with urea-formaldehyde (UF) resin and a novel ammonium lignosulfonate (ALS) is investigated in this paper. HDF panels were fabricated in the laboratory by applying a very low UF gluing factor (3%) and ALS content varying from 6% to 10% (based on the dry fibers). The physical and mechanical properties of the fiberboards, such as water absorption (WA), thickness swelling (TS), modulus of elasticity (MOE), bending strength (MOR), internal bond strength (IB), as well as formaldehyde content, were determined in accordance with the corresponding European standards. Overall, the HDF panels exhibited very satisfactory physical and mechanical properties, fully complying with the standard requirements of HDF for use in load-bearing applications in humid conditions. Markedly, the formaldehyde content of the laboratory fabricated panels was extremely low, ranging between 0.7–1.0 mg/100 g, which is, in fact, equivalent to the formaldehyde release of natural wood.

Keywords: wood-based panels; high-density fiberboards; bio-adhesives; ammonium lignosulfonate; zero-formaldehyde emission

check for **updates**

Citation: Antov, P.; Savov, V.; Krišťák, Ľ.; Réh, R.; Mantanis, G.I. Eco-Friendly, High-Density Fiberboards Bonded with Urea-Formaldehyde and Ammonium Lignosulfonate. *Polymers* **2021**, *13*, 220. https://doi.org/10.3390/polym13020220

Received: 17 December 2020
Accepted: 6 January 2021
Published: 10 January 2021

Publisher's Note: MDPI stays neutral with regard to jurisdictional claims in published maps and institutional affiliations.

1. Introduction

The growing need for sustainable products and the stringent legislative requirements related to the hazardous formaldehyde emissions from wood-based panels have boosted scientific and industrial interest in the production of eco-friendly, wood-based panels [1–7] and optimal utilization of the available lignocellulosic materials [8–13].

Fiberboards are wood-based panels produced by breaking down softwood and hardwood material into fibers, mixing them with wax and a formaldehyde-based thermosetting resin, such as urea-formaldehyde (UF), phenol formaldehyde, or melamine-urea formaldehyde, and forming panels by applying pressure and high temperature in a hot press [14,15]. Depending on the density, fiberboards can be classified into low-density fiberboards, with densities less than 400 kg·m^{-3}, medium-density fiberboards, with densities ranging from 400 to 900 kg·m^{-3}, and high-density fiberboards (HDF), with densities ranging from 900 to 1100 kg·m^{-3}. HDF panels are one of the most widely used wood-based products worldwide, with a variety of end-uses, such as high-grade furniture, laminate flooring, wall panels, shelves, and door skins. High-density fiberboards are denser than particleboards, and can be even denser than ordinary plywood; this characteristic broadens their application. HDF panels have many advantages, such as having a smoother surface, easier machinability, and increased strength. Resistance to direct screw withdrawal is relevant only for those fiberboard products used in furniture and cabinetmaking. Such panels are ideal for use as substrates for thin overlays in indoor conditions [16–18]. Another important advantage of HDF panels is the utilization of small-sized, low-quality wood, which is otherwise used mainly for energy purposes [19–21].

Traditional synthetic adhesives used in the production of wood-based panels are based on fossil-derived constituents, mainly urea, melamine, and phenol [22–25]. At present, about 95% of total wood adhesives used in the production of wood-based panels are formaldehyde-based resins [25]. UF resins dominate the global market, accounting for about 85% of the amino resins produced worldwide, with an estimated annual production of 11 million tons/year [2,25,26]. UF resins are thermosetting resins, the product of a reaction between urea and formaldehyde. These resins have been widely used in the production of engineered wood-based panels as a result of their numerous advantages, e.g., low press temperatures, short press times, excellent strength properties, chemical versatility, and a relatively low price [23–30]. The main drawback of UF resins, apart from their lower water resistance in comparison with phenol and melamine formaldehyde resins [31], is the release of hazardous volatile organic compounds (VOC) and free formaldehyde from the final UF-bonded products, especially with indoor use [32–34].

Formaldehyde (HCHO, CAS No. 50-00-0) is associated with a number of significant environmental issues [35] and serious human health hazards, such as skin and respiratory tract irritation, skin sensitization, nausea, genotoxicity, and cancer [36–38]. In 2004, formaldehyde was reclassified from "probable human carcinogen" to "known human carcinogen" (Group 1) by the International Agency for Research on Cancer [39]. Since then, the formaldehyde limit values have steadily been lowered, which has led to the development of less toxic, low-emission, eco-friendly, wood-based panels, where conventional synthetic resins have been partially or completely replaced by sustainable, bio-based adhesives [40–47] or by adding formaldehyde scavengers to adhesive formulations, such as urea [48], sodium metabisulfite ($Na_2S_2O_5$), ammonium bisulfite ((NH_4)HSO_3), other ammonium-containing agents [49,50], Al_2O_3 nanoparticles [51], etc. Various natural compounds, such as wood bark flours [32,52–54] and tannins [55–57], have also been successfully used to reduce the free formaldehyde content of wood panels.

Different renewable biomass resources, such as proteins [58–60], starch [61,62], tannins [63–65], and lignin [9,66–69] have been used as sustainable raw materials for the development of bio-based adhesives for engineered wood panels.

Lignin is a polyaromatic amorphous macromolecule and the second most abundant organic material after cellulose [70,71]. It is considered to be low-value waste or a byproduct, e.g., of the pulp and paper industries, which generate approximately 50–70 million tons of lignin annually [72]. With the increased production of biofuels, it is estimated that this number will reach 225 million tons per year by 2030 [72]. The enhanced scientific and industrial interest in lignin valorization is due to its abundance, renewability, and potential for providing an economically and environmentally sustainable alternative raw material for the production of value-added products, including adhesives for wood-based panels [4,43,73]. The application of lignin in the formulation of adhesives is primarily due to its polyphenolic structure. Thus, lignin may be used in lignin-phenol-formaldehyde resins, where it is applied to partially replace phenol [6,74,75]. Additional chemical modification of lignin, such as phenolation and methylolation, can be applied to increase the lignin reactivity to formaldehyde [1,75,76].

Technical lignins are obtained by different pulping technologies, but only lignosulfonates, which are water-soluble polyelectrolytes derived as byproducts of the sulfite processes, are available in large quantities, with an estimated annual production of approximately 800,000 tons [77–80]. Lignosulfonates contain a large number of functional groups, resulting in excellent surface activity; ammonium lignosulfonates (ALS) have been shown to be the most suitable ones for adhesive applications due to their large number of hydroxyl groups [75,81]. Another important advantage of ALS is their solubility in a number of organic solvents, unlike sodium, calcium, and magnesium salts, which are only soluble in water [82].

The aim of this research work is to investigate the potential of producing eco-friendly HDF panels from hardwood fibers, bonded with UF resin and a novel ALS adhesive, in order to reach the European standard requirements.

2. Materials and Methods

Industrially produced fibers obtained in factory conditions through the Asplund-method, using a L46 Defibrator (Valmet, Stockholm,, Sweden), were used in this work. The fibers were supplied by the factory of Welde Bulgaria AD (Troyan, Bulgaria). The fibers were obtained by the thermo-mechanical defibration of wood chips with dimensions (length) from 5 to 30 mm, subjected to steam treatment at 0.8 MPa steam pressure and 170 °C temperature. The industrial wood fibers had a bulk density of 29 $kg \cdot m^{-3}$ and a pulp freeness of 11° SR (Schopper Riegler test). The fibers were composed of the hardwood species European beech (*Fagus sylvatica* L.) and Turkey oak (*Quercus cerris* L.) at a ratio of 2:1, and were oven dried to a moisture content of 6.2%. The fibers had lengths from 1116 μm to 1250 μm (factory data).

Urea-formaldehyde resin with a solid content of 64% and a molar ratio (MR) of 1.16 was provided by Kastamonu Bulgaria AD (Gorno Sahrane, Bulgaria).

Commercial ammonium lignosulfonate (D-947L) was supplied by Borregaard (Sarpsborg, Norway). The ALS (CAS No. 8061-53-8) adhesive had the following characteristics: ammonium content—4.1%, sodium content—0.1%, total sulfur content—6.8%, high performance liquid chromatography (HPLC) sugars—20%, pH—4.5%, total solids content—48.6%, specific gravity—1.220, boiling point 104 °C, evaporation rate 0.4, vapor pressure 14.2 mmHg, and viscosity (cps)—400 at 25 °C.

Under laboratory conditions, the HDF panels were produced with dimensions of 400×400 mm^2, a thickness of 6 mm, and a target density of 910 $kg \cdot m^{-3}$. The adhesive formulation was comprised of UF resin at 3% and three addition levels of ALS (6%, 8% and 10%), based on the weight of fibers. The UF resin was used at 50% concentration. Urea at 3% based on the dry resin at 50% concentration was used as a formaldehyde scavenger. Ammonium sulfate (($NH_4)_2SO_4$, CAS No. CAS 7783-20-2) at 1.5% based on the dry UF resin and 2% based on dry D-947L at 30% concentration was used as a hardener.

A control panel was fabricated with 6% UF content, based on the dry fibers, and without ALS (HDF type 4). This resin addition level (6%) is typical for commercial HDF panels.

The manufacturing parameters of the laboratory-fabricated HDF panels are presented in Table 1.

Table 1. Manufacturing parameters used in this research work.

Panel No.	Adhesive Type	Target Density ($kg \cdot m^{-3}$)	UF Resin Content (%)	Ammonium Lignosulfonate Content (%)
1	UF + ALS	910	3	6
2	UF + ALS	910	3	8
3	UF + ALS	910	3	10
4	UF	910	6	0

The industrial wood fibers were mixed with the adhesive formulation in a high-speed laboratory glue blender (850 min^{-1}). The hot pressing process was carried out using a single opening hydraulic press (PMC ST 100, Italy). The press temperature used was 200 °C. The press factor applied was 30 $s \cdot mm^{-1}$. The following four-stage pressing regime was used: in the first stage, the pressure was increased to 4.5 MPa for 20 s; in the second stage, the pressure was steadily decreased to 1.2 MPa; in the third stage, the pressure was decreased to 0.6 MPa. The fourth pressing stage was performed at a pressure of 1.8 MPa. After pressing, the fabricated composites were conditioned for 10 days at 20 ± 2 °C and 65% relative humidity.

The physical and mechanical properties of the fabricated HDF panels (Figure 1) were tested according to European standards, namely EN 310, EN 317, EN 319, and EN 323 [83–86]. A precision laboratory balance Kern (Kern & Sohn GmbH, Balingen, Germany) with an accuracy of 0.01 g was used to determine the mass of the test specimens. The dimensions of the test pieces were measured using digital calipers with an accuracy of 0.01 mm. The physical properties (water absorption and thickness swelling) were measured after 24 h of immersion in water. The thickness swelling was assessed using the differences between the initial and final panel thicknesses, and the water absorption was determined using the difference in weight. The mechanical properties of the HDF panels were determined using a universal testing machine Zwick/Roell Z010 (Zwick/Roell GmbH, Ulm, Germany). For each parameter, eight HDF samples were used for testing.

Figure 1. High-density fiberboards from industrial hardwood fibers bonded with UF resin and ammonium lignosulfonate; 910 kg·m^{-3} target density, 6 mm thickness and three addition levels of ammonium lignosulfonate (6%, 8%, 10%).

The formaldehyde content of the laboratory-produced panels was tested in the laboratory of Kronospan Bulgaria EOOD (Veliko Tarnovo, Bulgaria) on four specimens in accordance with the commonly used Perforator method [87].

Variation and statistical analysis of the results was carried out by using the specialized software, QstatLab 6.0.

3. Results

3.1. Physical and Mechanical Properties

The results of the physical and mechanical properties of the HDF panels, comprised of industrial hardwood fibers bonded with UF resin and a novel ALS adhesive (D-947L), are presented in this part. The density of the laboratory-produced panels varied from 893 to 930 kg·m^{-3}, which was close to the targeted value. The differences in the final density of the panels were significantly below 5%; thus, it did not have a significant effect on the mechanical and physical properties.

The physical properties of the laboratory-produced HDF panels, i.e., water absorption (WA) and thickness swelling (TS), were determined after 24 h of immersion in water. Both WA and TS are critical panel properties that are directly correlated with the dimensional stability of wood-based panels [22,24].

A graphical representation of the WA (24 h) of the laboratory-produced HDF panels is presented in Figure 2.

It was determined that increasing the ALS content from 6% to 10% resulted in decreased WA values of HDF panels from 31.5% to 26.1%, respectively; this means an average improvement of this property by 21%. The increased ALS addition resulted in a gradual decrease of WA values—a relative improvement of the WA by 10%, while increasing the ALS content by 2%. ALS has a pH of about 4.5, and the increased addition of ALS in the UF resin resulted in a decreased pH of the adhesive mixture. This acidic condition resulted in decreased fiber moisture absorption, and hence, improved the water resistance of the finished HDF panels. The increase in resistance of the UF resin modified by ALS was also caused by the decreased brittleness of the adhesive, which, in the case of the unmodified UF, causes the cured resin to crack and allow moisture to penetrate into the bonded product [88,89]. Only the HDF panel, fabricated with 3% UF resin and 6% ALS content, had higher WA values compared with the control HDF panel, produced with 6% UF resin and without ALS addition. The HDF panels, produced with 8% ALS and 10% ALS, had 1.06 times and 1.16 times lower WA values than the control panel, respectively.

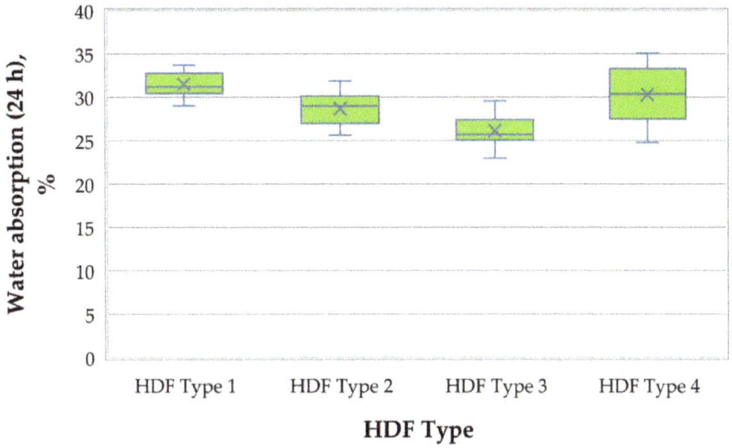

Figure 2. Water absorption (24 h) of HDF panels produced.

WA is not a standardized technical property; nonetheless, according to the literature [24,90], the WA of common HDF panels typically varies between 30% and 45%. Thus, the HDF panels, produced under laboratory conditions from industrial wood fibers, bonded with a UF resin and an eco-friendly, formaldehyde-free ALS adhesive, exhibited comparable or better WA values compared with the industrially-produced HDF panels bonded with formaldehyde-based adhesives.

A graphical representation of the TS (24 h) of the laboratory-produced HDF panels is shown in Figure 3.

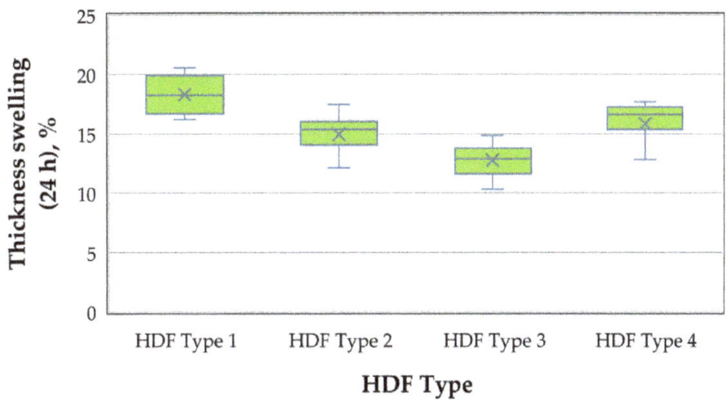

Figure 3. Thickness swelling of HDF panels produced.

As seen in Figure 3, TS of HDF panels bonded with ALS varied from 18.3% to 12.9%, i.e., increasing the ALS addition from 6% to 10% resulted in a 1.41 times improvement in TS values due to the decreased pH of the adhesive. All laboratory-produced panels had significantly better (lower) TS values than the standard requirement for application in humid conditions—25% [91]. The decreased pH of the adhesive may not be the only reason for the improvement in the TS values. Press temperature has a significant effect on the hygroscopic thickness swelling rate of HDF as well. The press temperature used in our research was 200 °C. It has been confirmed that the swelling rate increases as the HDF press temperature increases [92]. Only the HDF panel, produced with 3% UF resin and 6% ALS content, had higher TS values compared with the control panel (HDF type 4).

In terms of mechanical properties, the modulus of elasticity (MOE), bending strength (MOR), and internal bond (IB) strength of the laboratory-produced HDF panels were evaluated.

A graphic representation of the MOE of the laboratory-produced HDF panels is shown in Figure 4.

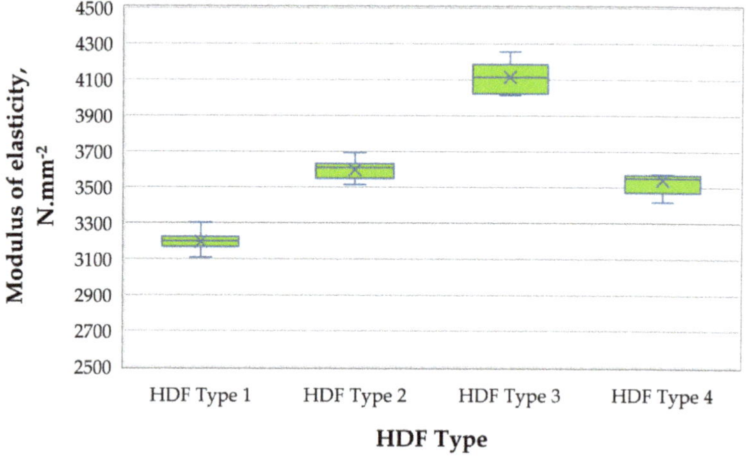

Figure 4. Modulus of elasticity (MOE) of HDF panels produced.

The MOE of fabricated HDF panels reached high values, ranging from 3197 to 4114 N·mm^{-2}. The estimated values significantly surpassed the European standard requirements [91] for HDF panels for use in humid conditions ($\geq$2900 N·mm^{-2}). Increasing the ALS addition from 6% to 10% resulted in a 29% improvement in MOE values. The incorporation of the increased ALS addition into the UF resin effectively reinforced the composites. The larger ALS quantity in the adhesive acted as a barrier against the influx of water, which resulted in better MOE values. Comparable results, i.e., MOE values ranging from 3730 to 4476 N·mm^{-2}, were reported by [6] in their work on the development of eco-friendly, medium-density fiberboards bonded with low phenol-formaldehyde (PF) resin content (3–5%) and calcium lignosulfonate, varying from 5% to 15%, depending on the dry fibers.

Only the HDF panel produced with 3% UF resin and 6% ALS addition content had lower MOE values compared with the control panel (HDF type 4). The HDF panel produced with 10% ALS content had 16% higher MOE values than the control panel.

A graphical representation of the MOR of the laboratory-produced HDF panels is shown in Figure 5.

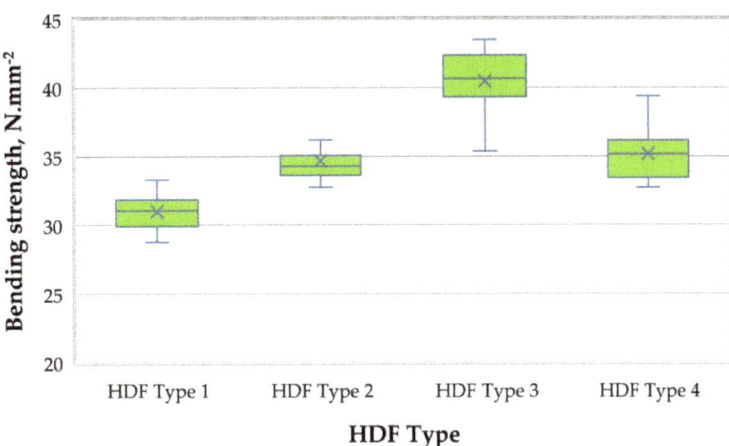

Figure 5. Bending strength (MOR) of HDF panels produced.

The fabricated HDF panels had very satisfactory MOR values, ranging from 30.99 N·mm^{-2} to 40.47 N·mm^{-2}, meeting the EN 622-2 standard requirements for HDF panels in humid conditions (MOR $\geq$ 30 N·mm^{-2}) [91]. MOR and MOR of HDF may be affected by a number of factors that are discussed in [93]. According to that study, the fiber dimensions have a significant effect on the physical, mechanical, and thermal properties of HDF panels. The wood species and digester conditions, i.e., press temperature, time, pressure, and defibrator grinding disc distance, are the most important parameters for the fiber quality. Increasing the ALS content from 6% to 10% resulted in improved MOR values by 31%. In the case of MOE, the greater ALS quantity in the adhesive mixture acted as a barrier against the water influx, which resulted in improved MOR values. The panels produced with 6% and 8% ALS content had lower bending strength values than the control panel. The HDF panel fabricated with 10% ALS content had a 27% greater MOR value than the control panel, bonded with 6% UF resin content. The maximum MOR value obtained in this work, i.e., 40.5 N·mm^{-2}, was determined at 3% UF resin content and 10% ALS addition. Similar values were reported by Antov et al. (2020) [6] in their work, in which the maximum MOR of 35.2 N·mm^{-2} was recorded for medium-density fiberboards bonded with 5% PF resin and 5% calcium lignosulfonate.

Finally, a graphical representation of the average IB strength of the laboratory-produced HDF panels is shown in Figure 6.

The IB strength of the HDF panels ranged from 0.58 N·mm^{-2} to 0.67 N·mm^{-2}; this means that increasing the ALS content from 6% to 10% resulted in a 17% increase in IB strength values. A significant improvement, i.e., by 12%, was recorded when the ALS content was increased from 6% to 8%. An increase in ALS significantly improved the interfacial compatibility of the materials. ALS acts as an anionic surfactant, and its molecular structure contains not only polar groups (e.g., hydroxyl and sulfonic groups), but also nonpolar groups (e.g., benzene propane skeleton and aliphatic side chains). With the increase of ALS, its activation gradually reduced the interfacial tension and interfacial free energy, thereby improving the mechanical properties of the composites [94]. There is an apparent correlation between internal bond and thickness swelling; the better a HDF is bonded, the better it resists the forces trying to cause thickness swelling. All laboratory-produced HDF panels met the standard requirements for HDF use in dry conditions (IB $\geq$ 0.50 N·mm^{-2}) [91]. Only the HDF panel bonded with 3% UF resin content and 6% ALS had a lower IB value than the control panel. The laboratory-produced HDF panel, produced with 10% ALS content, had a 13% greater IB strength value compared with the control panel.

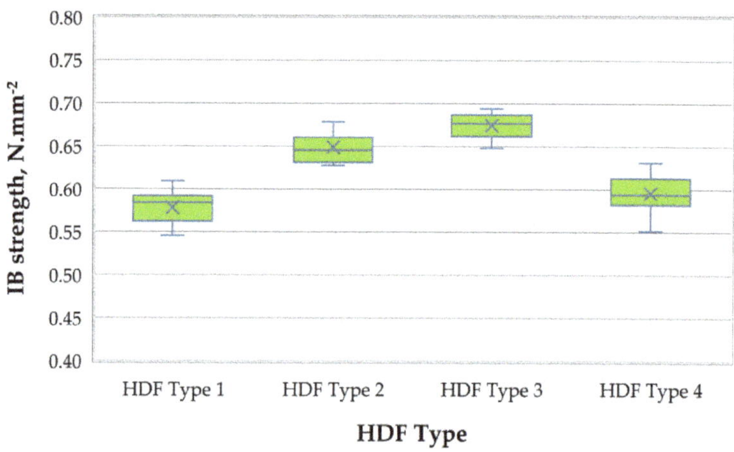

Figure 6. IB strength of HDF panels produced.

3.2. Formaldehyde Content

The results for the free formaldehyde content of the fabricated HDF panels, tested in accordance with the standard EN ISO 12460-5 (called the Perforator method), are presented in Table 2.

Table 2. Formaldehyde content of HDF panels according to EN ISO 12460-5.

HDF Type	Adhesive	UF Resin Content (%)	Ammonium Lignosulfonate Content (%)	Formaldehyde Content (mg/100 g)
1	UF + ALS	3	6	1.0 ± 0.1
2	UF + ALS	3	8	0.8 ± 0.1
3	UF + ALS	3	10	0.7 ± 0.1
4	UF	6	-	4.3 ± 0.1

The results obtained for the free formaldehyde content of the laboratory-produced HDF panels from industrial hardwood fibers, bonded with UF resin and ammonium lignosulfonate adhesive (D-947L), were remarkably low and can be considered as a zero-formaldehyde content [1,24]. All laboratory-fabricated HDF panels fulfilled the requirements of the super E0 emission grade ($\leq$1.5 mg/100 g). The lowest formaldehyde content of 0.7 ± 0.1 mg/100 g was achieved for the HDF panel bonded with 3% UF resin and 10% ALS (D-947L) content. In accordance with the free formaldehyde content results, the reference HDF panel, bonded with 6% UF resin only, can be classified under the emission class E1 ($\leq$8 mg/100 g). These results are in agreement with previous research, where using lignosulfonates as binders for wood composites resulted in decreased free formaldehyde content [6,9,45,69]. ALS has very good characteristics for methylolation due to its large number of phenolic hydroxyl groups and amount of aromatic protons of the guaiacyl units, whose presence tends to increase the reactivity of lignosulfonate toward formaldehyde [94,95].

Natural wood releases low, yet still measurable, amounts of formaldehyde formed by its main polymeric components and extractives at approximately 0.5 to 2 mg/100 g [96–98]. Considering this, HDF panels bonded with UF resin and ALS (D-947L) can be defined as extremely low-emission wood-based panels.

4. Conclusions

Eco-friendly HDF panels with acceptable physical-mechanical properties and close-to-zero formaldehyde emissions, fulfilling the European standards, can be produced from hardwood fibers bonded with a very low conventional UF resin (3%) and a novel ammonium lignosulfonate at a content of 6% to 10%, depending on the dry fibers. The laboratory-fabricated HDF panels met the stringent standard requirements for use in load-bearing applications in humid conditions. The formaldehyde content of panels produced in the laboratory was distinctly low, ranging from 0.7 mg/100 g to 1.0 mg/100 g (according to EN 12460-5), which is equivalent to the formaldehyde release of natural wood. The HDF panels manufactured with 3% UF gluing content and ammonium lignosulfonate addition >8% exhibited superior physical and mechanical properties compared with those of the control panels produced with a straight UF resin (at 6%). Future studies should focus on decreasing the hot-pressing factor by modifying the formula of ammonium lignosulfonate by adding suitable cross-linking agents, and studying in-depth the bonding interaction among formaldehyde-based resin, lignosulfonate additives, and lignocellulosic fibers.

Author Contributions: Conceptualization, P.A., R.R. and G.I.M.; methodology, P.A., V.S., R.R. and G.I.M.; validation, P.A., L'.K., R.R., and G.I.M.; investigation, P.A., V.S. and L'.K.; resources, P.A. and V.S.; writing—original draft preparation, P.A., L'.K., R.R. and G.I.M.; writing—review and editing, P.A., L'.K. and G.I.M.; visualization, P.A. and V.S.; supervision, R.R.; project administration, R.R.; funding acquisition, R.R. All authors have read and agreed to the published version of the manuscript.

Funding: This research was supported by the Slovak Research and Development Agency under contracts no. APVV-18-0378, APVV-19-0269, APVV-17-583 and VEGA1/0717/19.

Institutional Review Board Statement: Not applicable.

Informed Consent Statement: Not applicable.

Data Availability Statement: Data sharing is not applicable to this article.

Acknowledgments: The authors would like to thank: (i) Borregaard (Sarpsborg, Norway) for kindly providing the ammonium lignosulfonate used; (ii) Welde Bulgaria AD (Troyan, Bulgaria) for supplying the industrial wood fibers; (iii) Kastamonu Bulgaria AD (Gorno Sahrane, Bulgaria) for providing the urea-formaldehyde resin; and (iv) Kronospan Bulgaria EOOD (Veliko Tarnovo, Bulgaria) for determining the formaldehyde content of the HDF specimens. This publication is also the result of the following project implementation: Progressive research of performance properties of wood-based materials and products (LignoPro), ITMS 313011T720 supported by the Operational Programme Integrated Infrastructure (OPII) funded by the ERDF.

Conflicts of Interest: The authors declare no conflict of interest.

References

1. Pizzi, A. Recent developments in eco-efficient bio-based adhesives for wood bonding: Opportunities and issues. *J. Adhes. Sci. Technol.* **2006**, *20*, 829–846. [CrossRef]
2. Pizzi, A.; Papadopoulos, A.; Policardi, F. Wood Composites and Their Polymer Binders. *Polymers* **2020**, *12*, 1115. [CrossRef] [PubMed]
3. Taghiyari, H.R.; Hosseini, S.B.; Ghahri, S.; Ghofrani, M.; Papadopoulos, A.N. Formaldehyde Emission in Micron-Sized Wollastonite-Treated Plywood Bonded with Soy Flour and Urea-Formaldehyde Resin. *Appl. Sci.* **2020**, *10*, 6709. [CrossRef]
4. Ferdosian, F.; Pan, Z.; Gao, G.; Zhao, B. Bio-based adhesives and evaluation for wood composites application. *Polymers* **2017**, *9*, 70. [CrossRef]
5. Taghiyari, H.R.; Esmailpour, A.; Majidi, R.; Morrell, J.J.; Mallaki, M.; Militz, H.; Papadopoulos, A.N. Potential Use of Wollastonite as a Filler in UF Resin Based Medium-Density Fiberboard (MDF). *Polymers* **2020**, *12*, 1435. [CrossRef]
6. Antov, P.; Savov, V.; Mantanis, G.I.; Neykov, N. Medium-density fibreboards bonded with phenol-formaldehyde resin and calcium lignosulfonate as an eco-friendly additive. *Wood Mater. Sci. Eng.* **2020**, 1751279. [CrossRef]
7. Taghiyari, H.R.; Tajvidi, M.; Taghiyari, R.; Mantanis, G.I.; Esmailpour, A.; Hosseinpourpia, R. Nanotechnology for Wood Quality Improvement and Protection. In *Nanomaterials for Agriculture and Forestry Applications*; Husen, A., Jawaid, M., Eds.; Elsevier: Amsterdam, The Netherlands, 2020; pp. 469–489.
8. Ihnát, V.; Lübke, H.; Balberčák, J.; Kuňa, V. Size Reduction Downcycling of Waste Wood—A Review. *Wood Res.* **2020**, *65*, 205–220. [CrossRef]
9. Antov, P.; Mantanis, G.I.; Savov, V. Development of wood composites from recycled fibres bonded with magnesium lignosulfonate. *Forests* **2020**, *11*, 613. [CrossRef]
10. Iždinský, J.; Vidholdová, Z.; Reinprecht, L. Particleboards from Recycled Wood. *Forests* **2020**, *11*, 1166. [CrossRef]

11. Ihnát, V.; Lübke, H.; Russ, A.; Pažitný, A.; Borůvka, V. Waste agglomerated wood materials as a secondary raw material for chipboards and fibreboards. Part II: Preparation and characterization of wood fibres in terms of their reuse. *Wood Res.* **2018**, *63*, 431–442.
12. Mantanis, G.; Athanassiadou, E.; Nakos, P.; Coutinho, A. A New Process for Recycling Waste Fiberboards. In Proceedings of the 38th International Wood Composites Symposium, Washington, DC, USA, 5–8 April 2004; pp. 119–122.
13. Barbu, M.C.; Sepperer, T.; Tudor, E.M.; Petutschnigg, A. Walnut and Hazelnut Shells: Untapped Industrial Resources and Their Suitability in Lignocellulosic Composites. *Appl. Sci.* **2020**, *10*, 6340. [CrossRef]
14. Kúdela, J. Surface properties of a medium density fibreboard evaluated from the viewpoint of its surface treatment. *Acta Facultatis Xylologiae Zvolen* **2020**, *62*, 35–45.
15. Lubis, M.A.R.; Hong, M.K.; Park, B.D.; Lee, S.M. Effects of recycled fiber content on the properties of medium density fiberboard. *Eur. J. Wood Prod.* **2018**, *76*, 1515–1526. [CrossRef]
16. Zhu, Z.; Buck, D.; Guo, X.; Ekevad, M.; Cao, P.; Wu, Z. Machinability investigation in turning of high density fiberboard. *PLoS ONE* **2018**, *13*, e0203838. [CrossRef] [PubMed]
17. Kovatchev, G. Influence of the Belt Type Over Vibration of the Cutting Mechanism in Woodworking Shaper. In Proceedings of the 11th International Science Conference "Chip and Chipless Woodworking Processes", Zvolen, Slovakia, 13–15 September 2018; pp. 105–110.
18. Wagner, K.; Schnabel, T.; Barbu, M.C.; Petutschnigg, A. Analysis of selected properties of fiberboard panels manufactured from wood and leather using the near infrared spectroscopy. *Int. J. Spectrosc.* **2015**, *2015*, 691796. [CrossRef]
19. Tritchkov, N.; Antov, P. Prospects for Developing the Production of Solid Wood Products Taking into Account the Raw-Material Base. In Proceedings of the COST Action E44 Conference "Broad Spectrum Utilisation of Wood", Vienna, Austria, 14–15 June 2005; pp. 131–139.
20. Kulman, S.; Boiko, L.; Gurová, D.H.; Sedliačik, J. The Effect of Temperature and Moisture Changes on Modulus of Elasticity and Modulus of Rupture of Particleboard. *Acta Facultatis Xylologiae Zvolen* **2019**, *61*, 43–52.
21. Savov, V.; Antov, P. Engineering the Properties of Eco-Friendly Medium Density Fibreboards Bonded with Lignosulfonate Adhesive. *Drvna Industrija* **2020**, *71*, 157–162. [CrossRef]
22. Youngquist, J.A. Wood-Based Composites and Panel Products. In *Wood Handbook: Wood as an Engineering Material*; USDA Forest Service, Forest Products Laboratory: Madison, WI, USA, 1999; pp. 1–31.
23. Wibowo, E.S.; Lubis, M.A.R.; Park, B.D.; Kim, J.S.; Causin, V. Converting crystalline thermosetting urea-formaldehyde resins to amorphous polymer using modified nanoclay. *J. Ind. Eng. Chem.* **2020**, *87*, 78–89. [CrossRef]
24. Mantanis, G.I.; Athanassiadou, E.T.; Barbu, M.C.; Wijnendaele, K. Adhesive systems used in the European particleboard, MDF and OSB industries. *Wood Mater. Sci. Eng.* **2018**, *13*, 104–116. [CrossRef]
25. Kumar, R.N.; Pizzi, A. Environmental Aspects of Adhesives–Emission of Formaldehyde. In *Adhesives for Wood and Lignocellulosic Materials*; Wiley-Scrivener Publishing: Hoboken, NJ, USA, 2019; pp. 293–312.
26. Dae, P.B.; Woo, K.J. Dynamic mechanical analysis of urea-formaldehyde resin adhesives with different formaldehyde-to-urea molar ratios. *J. Appl. Polym. Sci.* **2008**, *108*, 2045–2051.
27. Bekhta, P.; Sedliacik, J.; Saldan, R.; Novak, I. Effect of different hardeners for urea-formaldehyde resin on properties of birch plywood. *Acta Fac. Xylologiae Zvolen* **2016**, *58*, 65–72.
28. Dunky, M. Adhesives in the Wood Industry. In *Handbook of Adhesive Technology*; Pizzi, A., Mittal, K.L., Eds.; Marcel Dekker: New York, NY, USA, 2003; Chapter 47; pp. 872–941.
29. Zhang, W.; Ma, Y.; Wang, C.; Li, S.; Zhang, M.; Chu, F. Preparation and properties of lignin-phenol-formaldehyde resins based on different biorefinery residues of agricultural biomass. *Ind. Crops Prod.* **2013**, *43*, 326–333. [CrossRef]
30. Jivkov, V.; Simeonova, R.; Marinova, A.; Gradeva, G. Study on the Gluing Abilities of Solid Surface Composites with Different Wood Based Materials and Foamed PVC. In Proceedings of the 24th International Scientific Conference Wood Is Good–User Oriented Material, Technology and Design, Zagreb, Croatia, 18 October 2013; pp. 49–55, ISBN 978-953-292-031-4.
31. Kumar, R.N.; Pizzi, A. Urea-Formaldehyde Resins. In *Adhesives for Wood and Lignocellulosic Materials*; Wiley-Scrivener Publishing: Hoboken, NJ, USA, 2019; pp. 61–100.
32. Tudor, E.M.; Barbu, M.C.; Petutschnigg, A.; Réh, R.; Krišťák, Ľ. Analysis of Larch-Bark Capacity for Formaldehyde Removal in Wood Adhesives. *Int. J. Environ. Res. Public Health* **2020**, *17*, 764. [CrossRef] [PubMed]
33. Tudor, E.M.; Dettendorfer, A.; Kain, G.; Barbu, M.C.; Réh, R.; Krišťák, Ľ. Sound-Absorption Coefficient of Bark-Based Insulation Panels. *Polymers* **2020**, *12*, 1012. [CrossRef]
34. Mirski, R.; Bekhta, P.; Dziurka, D. Relationships between Thermoplastic Type and Properties of Polymer-Triticale Boards. *Polymers* **2019**, *11*, 1750. [CrossRef]
35. Łebkowska, M.; Radziwiłł, M.Z.; Tabernacka, A. Adhesives based on formaldehyde—Environmental problems. *BioTechnologia* **2017**, *98*, 53–65. [CrossRef]
36. U.S. Consumer Product Safety Commission. *An Update on Formaldehyde (Publication 725)*; U.S. Consumer Product Safety Commission: Bethesda, MD, USA, 2013.
37. Bekhta, P.; Sedliačik, J.; Noshchenko, G.; Kačík, F.; Bekhta, N. Characteristics of Beech Bark and its Effect on Properties of UF Adhesive and on Bonding Strength and Formaldehyde Emission of Plywood Panels. *Eur. J. Wood Prod.* **2021**. [CrossRef]

38. World Health Organization. Formaldehyde, 2–Butoxyethanol and 1–tert–Butoxypropan–2–ol. In *Monographs on the Evaluation of Carcinogenic Risk to Humans*; International Agency for Research on Cancer: Lyon, France, 2006; Volume 88.

39. International Agency for Research on Cancer. *IARC Classifies Formaldehyde as Carcinogenic to Humans*; International Agency for Research on Cancer: Lyon, France, 2004.

40. Kawalerczyk, J.; Siuda, J.; Mirski, R.; Dziurka, D. Hemp Flour as a Formaldehyde Scavenger for Melamine-Urea-Formaldehyde Adhesive in Plywood Production. *Bioresources* **2020**, *15*, 4052–4064.

41. Papadopoulou, E. Adhesives from renewable resources for binding wood-based panels. *J. Environ. Prot. Ecol.* **2009**, *10*, 1128–1136.

42. Nordström, E.; Demircan, D.; Fogelström, L.; Khabbaz, F.; Malmström, E. Green Binders for Wood Adhesives. In *Applied Adhesive Bonding in Science and Technology*; Interhopen Books: London, UK, 2017; pp. 47–71.

43. Hemmilä, V.; Adamopoulos, S.; Karlsson, O.; Kumar, A. Development of sustainable bio-adhesives for engineered wood panels-A review. *RSC Adv.* **2017**, *7*, 38604–38630. [CrossRef]

44. Hosseinpourpia, R.; Adamopoulos, S.; Mai, C.; Taghiyari, H.R. Properties of medium-density fiberboards bonded with dextrin-based wood adhesives. *Wood Res.* **2019**, *64*, 185–194.

45. Antov, P.; Savov, V.; Neykov, N. Sustainable Bio-based Adhesives for Eco-Friendly Wood Composites—A Review. *Wood Res.* **2020**, *65*, 51–62. [CrossRef]

46. Dunky, M. *Wood Adhesives Based on Natural Resources: A Critical Review Part II. Carbohydrate-Based Adhesives, In Reviews of Adhesion and Adhesives*; Scrivener Publishing: Hoboken, NJ, USA, 2020; Volume 8, pp. 333–378.

47. Sarika, P.R.; Nancarrow, P.; Khansaheb, A.; Ibrahim, T. Bio-Based Alternatives to Phenol and Formaldehyde for the Production of Resins. *Polymers* **2020**, *12*, 2237. [CrossRef] [PubMed]

48. Park, B.D.; Kang, E.C.; Park, J.Y. Thermal curing behavior of modified urea-formaldehyde resin adhesives with two formaldehyde scavengers and their influence on adhesion performance. *J. Appl. Polym. Sci.* **2008**, *110*, 1573–1580. [CrossRef]

49. Costa, N.; Pereira, J.; Ferra, J.; Cruz, P.; Martins, J.; Magalhães, F.; Mendes, A.; Carvalho, L.H. Scavengers for achieving zero formaldehyde emission of wood-based panels. *Wood Sci. Technol.* **2013**, *47*, 1261–1272. [CrossRef]

50. Costa, N.; Pereira, J.; Martins, J.; Ferra, J.; Cruz, P.; Magalhães, F.; Mendes, A.; Carvalho, L. Alternative to latent catalysts for curing UF resins used in the production of low formaldehyde emission wood-based panels. *Int. J. Adhes. Adhes.* **2012**, *33*, 56–60. [CrossRef]

51. de Cademartori, P.H.G.; Artner, M.A.; de Freitas, R.A.; Magalhaes, W.L.E. Alumina nanoparticles as formaldehyde scavenger for urea-formaldehyde resin: Rheological and in-situ cure performance. *Compos. B Eng.* **2019**, *176*, 107281. [CrossRef]

52. Medved, S.; Gajsek, U.; Tudor, E.M.; Barbu, M.C.; Antonovic, A. Efficiency of bark for reduction of formaldehyde emission from particleboards. *Wood Res.* **2019**, *64*, 307–315.

53. Réh, R.; Igaz, R.; Krišt'ák, L'.; Ružiak, I.; Gajtanska, M.; Božíková, M.; Kučerka, M. Functionality of beech bark in adhesive mixtures used in plywood and its effect on the stability associated with material systems. *Materials* **2019**, *12*, 1298. [CrossRef]

54. Mirski, R.; Kawalerczyk, J.; Dziurka, D.; Siuda, J.; Wieruszewski, M. The Application of Oak Bark Powder as a Filler for Melamine-Urea-Formaldehyde Adhesive in Plywood Manufacturing. *Forests* **2020**, *11*, 1249. [CrossRef]

55. Boran, S.; Usta, M.; Ondaral, S.; Gümüskaya, E. The efficiency of tannin as a formaldehyde scavenger chemical in medium density fiberboard. *Compos. B Eng.* **2012**, *43*, 2487–2491. [CrossRef]

56. Bekhta, P.; Sedliacik, J.; Kacik, F.; Noshchenko, G.; Kleinova, A. Lignocellulosic waste fibers and their application as a component of urea-formaldehyde adhesive composition in the manufacture of plywood. *Eur. J. Wood Prod.* **2019**, *77*, 495–508. [CrossRef]

57. Antov, P.; Savov, V.; Neykov, N. Reduction of Formaldehyde Emission from Engineered Wood Panels by Formaldehyde Scavengers—A Review. In Proceedings of the 13th International Scientific Conference Wood EMA 2020 and 31st International Scientific Conference ICWST 2020 Sustainability of Forest-Based Industries in the Global Economy, Vinkovci, Croatia, 28–30 September 2020; pp. 7–11, ISBN 978-953-57822-8-5.

58. Wang, Z.; Zhao, S.; Pang, H.; Zhang, W.; Zhang, S.; Li, J. Developing eco-friendly high-strength soy adhesives with improved ductility through multiphase core–shell hyperbranchedpolysiloxane. *ACS Sustain. Chem. Eng.* **2019**, *7*, 7784–7794. [CrossRef]

59. Frihart, C.; Birkeland, M. *Soy Properties and Soy Wood Adhesives*; American Chemical Society: Midland, MI, USA, 2014; pp. 167–192.

60. Zhang, B.; Zhang, F.; Wu, L.; Gao, Z.; Zhang, L. Assessment of soybean protein-based adhesive formulations, prepared by different liquefaction technologies for particleboard applications. *Wood Sci. Technol.* **2020**. [CrossRef]

61. Tan, H.; Zhang, Y.; Weng, X. Preparation of the Plywood Using Starch-based Adhesives Modified with blocked isocyanates. *Procedia Eng.* **2011**, *15*, 1171–1175. [CrossRef]

62. Li, Z.; Wang, J.; Li, C.; Gu, Z.; Cheng, L.; Hong, Y. Effects of montmorillonite addition on the performance of starch-based wood adhesive. *Carbohydr. Polym.* **2015**, *115*, 394–400. [CrossRef]

63. Ndiwe, B.; Pizzi, A.; Tibi, B.; Danwe, R.; Konai, N.; Amirou, S. African tree bark exudate extracts as biohardeners of fully biosourced thermoset tannin adhesives for wood panels. *Ind. Crops Prod.* **2019**, *132*, 253–268. [CrossRef]

64. Santos, J.; Antorrena, G.; Freire, M.S.; Pizzi, A.; Álvarez, J.G. Environmentally friendly Wood adhesives based on chestnut (*Castanea sativa*) shell tannins. *Eur. J. Wood Wood Prod.* **2017**, *75*, 89–100. [CrossRef]

65. Konai, N.; Pizzi, A.; Danwe, R.; Lucien, M.; Lionel, K.T. Thermomechanical analysis of African tannins resins and biocomposite characterization. *J. Adhes. Sci. Technol.* **2020**, 1850611. [CrossRef]

66. El Mansouri, N.E.; Pizzi, A.; Salvadó, J. Lignin-based wood panel adhesives without formaldehyde. *Holz Roh Werkst.* **2006**, *65*, 65. [CrossRef]

67. Li, R.J.; Gutierrez, J.; Chung, Y.; Frank, C.W.; Billington, S.L.; Sattely, E.S. A lignin-epoxy resin derived from biomass as an alternative to formaldehyde-based wood adhesives. *Green Chem.* **2018**, *20*, 1459–1466. [CrossRef]
68. Gadhave, R.V.; Srivastava, S.; Mahanwar, P.A.; Gadekar, P.T. Lignin: Renewable Raw Material for Adhesive. *Open J. Polym. Chem.* **2019**, *9*, 27–38. [CrossRef]
69. Antov, P.; Jivkov, V.; Savov, V.; Simeonova, R.; Yavorov, N. Structural Application of Eco-Friendly Composites from Recycled Wood Fibres Bonded with Magnesium Lignosulfonate. *Appl. Sci.* **2020**, *10*, 7526. [CrossRef]
70. Lora, J.H.; Glasser, W.G. Recent industrial applications of lignin: A sustainable alternative to nonrenewable materials. *J. Polym. Environ.* **2002**, *10*, 39–48. [CrossRef]
71. *Lignin: Biosynthesis and Transformation for Industrial Applications*; Sharma, S.; Kumar, A. (Eds.) Springer Series on Polymer and Composite Materials, Switzerland AG; Springer Nature: Berlin/Heidelberg, Germany, 2020.
72. Bajwa, D.S.; Pourhashem, G.; Ullah, A.H.; Bajwa, S.G. A concise review of current lignin production, applications, products and their environmental impact. *Ind. Crop Prod.* **2019**, *139*, 111526. [CrossRef]
73. Klapiszewski, Ł.; Oliwa, R.; Oleksy, M.; Jesionowski, T. Calcium lignosulfonate as eco-friendly additive of crosslinking fibrous composites with phenol-formaldehyde resin matrix. *Polymers* **2018**, *63*, 102–108. [CrossRef]
74. Jin, Y.; Cheng, X.; Zheng, Z. Preparation and characterization of phenol–formaldehyde adhesives modified with enzymatic hydrolysis lignin. *Bioresour. Technol.* **2010**, *101*, 2046–2048. [CrossRef]
75. Hemmilä, V.; Adamopoulos, S.; Hosseinpourpia, R.; Sheikh, A.A. Ammonium lignosulfonate adhesives for particleboards with pMDI and furfuryl alcohol as cross-linkers. *Polymers* **2019**, *11*, 1633. [CrossRef]
76. Vázquez, G.; González, J.; Freire, S.; Antorrena, G. Effect of chemical modification of lignin on the gluebond performance of lignin-phenolic resins. *Bioresour. Technol.* **1997**, *60*, 191–198. [CrossRef]
77. Lora, J. Industrial Commercial Lignins: Sources, Properties and Applications. In *Monomers, Polymers and Composites from Renewable Resources*; Belgacem, M.N., Gandini, A., Eds.; Elsevier: Amsterdam, The Netherlands, 2008; pp. 225–241.
78. Agrawal, A.; Kaushik, N.; Biswas, S. Derivatives and applications of lignin—An insight. *SciTech J.* **2014**, *1*, 30–36.
79. Berlin, A.; Balakshin, M. Industrial Lignins: Analysis, Properties, and Applications. In *Bioenergy Research: Advances and Applications*; Gupta, V.K., Tuohy, M., Kubicek, C., Saddler, J., Xu, F., Eds.; Elsevier: Amsterdam, The Netherlands, 2014; Chapter 18; pp. 315–336.
80. Vishtal, A.G.; Kraslawski, A. Challenges in industrial applications of technical lignins. *BioResources* **2011**, *6*, 3547–3568.
81. Alonso, M.V.; Oliet, M.; Rodríguez, F.; Astarloa, G.; Echeverría, J.M. Use of a methylolated softwood ammonium lignosulfonate as partial substitute of phenol in resol resins manufacture. *J. Appl. Polym. Sci.* **2004**, *94*, 643–650. [CrossRef]
82. Alonso, M.V.; Oliet, M.; Rodríguez, F.; García, J.; Gilarranz, M.A.; Rodríguez, J.J. Modification of ammonium lignosulfonate by phenolation for use in phenolic resins. *Bioresour. Technol.* **2005**, *96*, 1013–1018. [CrossRef] [PubMed]
83. EN 310. *Wood-Based Panels-Determination of Modulus of Elasticity in Bending and of Bending Strength*; European Committee for Standardization: Brussels, Belgium, 1999.
84. EN 317. *Particleboards and Fibreboards-Determination of Swelling in Thickness after Immersion in Water*; European Committee for Standardization: Brussels, Belgium, 1998.
85. EN 322. *Wood-Based Panels-Determination of Moisture Content*; European Committee for Standardization: Brussels, Belgium, 1998.
86. EN 323. *Wood-Based Panels-Determination of Density*; European Committee for Standardization: Brussels, Belgium, 2001.
87. EN ISO 12460-5. *Wood-Based Panels-Determination of Formaldehyde Release—Part 5. Extraction Method (Called the Perforator Method)*; European Committee for Standardization: Brussels, Belgium, 2015.
88. Kordkheili, H.Y.; Najafi, S.K.; Eshkiki, R.B.; Pizzi, A. Improving urea formaldehyde resin properties by glyoxalated soda bagasse lignin. *Eur. J. Wood. Prod.* **2015**, *73*, 77–85. [CrossRef]
89. Kim, J.W.; Carlbom, K.; Matuana, L.; Heiden, P. Thermoplastic modification of urea-formaldehyde wood adhesives to improve moisture resistance. *J. Appl. Polym. Sci.* **2006**, *101*, 4222–4229. [CrossRef]
90. Mihajlova, J.; Savov, V. Physical Indicators of High-Density Fibreboards (HDF) Manufactured from Wood of Hard Broadleaved Species. In Proceedings of the 8th Hardwood Conference, Sopron, Hungary, 25–26 October 2018; pp. 142–144, ISBN 978-963-359-096-6.
91. EN 622-2. *Fibreboards–Specifications—Part 2: Requirements for Hardboard*; European Committee for Standardization: Brussels, Belgium, 2004.
92. Shi, S.Q.; Gardner, D.J. Hygroscopic thickness swelling rate of compression molded wood fiberboard and wood fiber/polymer composites. *Compos. Part A Appl. Sci. Manuf.* **2006**, *37*, 1276–1285. [CrossRef]
93. Nasir, M.; Khali, D.P.; Jawaid, M.; Tahir, P.M.; Siakeng, R.; Asim, M.; Khan, T.A. Recent development in binderless fiber-board fabrication from agricultural residues: A review. *Constr. Build. Mater.* **2019**, *211*, 502–516. [CrossRef]
94. Hu, J.P.; Guo, M.H. Influence of ammonium lignosulfonate on the mechanical and dimensional properties of wood fiber biocomposites reinforced with polyactic acid. *Ind. Crops Prod.* **2015**, *78*, 48–57. [CrossRef]
95. Alonso, M.V.; Rodriguez, J.J.; Oliet, M.; Rodriguez, F.; Garcia, J.; Gilarranz, M.A. Characterization and Structural Modification of Ammonic Lignosulfonate by Methylolation. *J. Appl. Polym. Sci.* **2001**, *82*, 2661–2668. [CrossRef]
96. Roffael, E. Volatile organic compounds and formaldehyde in nature, wood and wood based panels. *Holz Roh Werkst.* **2006**, *64*, 144–149. [CrossRef]

97. Athanassiadou, E.; Roffael, E.; Mantanis, G. Medium Density Fiberboards (MDF) from Recycled Fibres. In Proceedings of the Conference "Towards a Higher Technical, Economical and Environmental Standard in Europe" COST Action E31, Bordeaux, France, 29 September–1 October 2005; pp. 248–261.
98. Salem, M.Z.M.; Böhm, M. Understanding of formaldehyde emissions from solid wood: An overview. *BioResources* **2013**, *8*, 4775–4790. [CrossRef]

 *polymers*

Article

The Impact of Wood Waste on the Properties of Silicone-Based Composites

Maciej Mrówka [1], Małgorzata Szymiczek [1,*] and Magdalena Skonieczna [2,3]

1   Department of Theoretical and Applied Mechanics, Silesian University of Technology, Konarskiego 18 A,
    44-100 Gliwice, Poland; Maciej.Mrowka@polsl.pl
2   Department of Systems Biology and Engineering, Silesian University of Technology, Akademicka 16,
    44-100 Gliwice, Poland; Magdalena.Skonieczna@polsl.pl
3   Biotechnology Centre, Silesian University of Technology, Krzywoustego 8, 44-100 Gliwice, Poland
*   Correspondence: Malgorzata.Szymiczek@polsl.pl; Tel.: +48-32-237-12-43

Abstract: The impact of wood waste on the mechanical and biological properties of silicone-based composites was investigated using wood waste from oak, hornbeam, beech, and spruce trees. The density, abrasion resistance, resilience, hardness, and static tensile properties of the obtained WPC (wood–plastic composites) were tested. The results revealed slight changes in the density, increased abrasion resistance, decreased resilience, increased hardness, and decreased strain at break and stress at break compared with untreated silicone. The samples also showed no cytotoxicity to normal human dermal fibroblast, NHDF. The possibility of using prepared composites as materials to create structures on the seabed was also investigated by placing samples in a marine aquarium for one week and then observing sea algae growth.

Keywords: wood–plastic composite (WPC); silicone; mechanical properties; cytotoxicity; casting; ageing

Citation: Mrówka, M.; Szymiczek, M.; Skonieczna, M. The Impact of Wood Waste on the Properties of Silicone-Based Composites. *Polymers* **2021**, *13*, 7. https://dx.doi.org/10.3390/polym13010007

Received: 30 November 2020
Accepted: 17 December 2020
Published: 22 December 2020

Publisher's Note: MDPI stays neutral with regard to jurisdictional claims in published maps and institutional affiliations.

1. Introduction

Polymeric fillers are very popular and include natural fillers, which can be derived from both vegetables (maize husk, nutshell, ground coffee, bran, and starch) and animals (milled bird feathers). Such fillers are introduced primarily into thermoplastic polymer materials [1–4]. An important feature of this type of material, apart from the ability to change their physical and mechanical properties, is the ability to modify their environmental properties. Wood–plastic composite (WPC) with a thermoplastic matrix, which are mainly obtained by extrusion and injection, have found applications in the construction industry (boards, furniture, and lagging) [2,4–6]. The introduction of biofillers for duroplasts, along with changes in the physical and mechanical properties, allows resin shrinkage to be minimized, but it also increases the resin viscosity, making it difficult to saturate fabrics. A separate issue is the introduction of silicone biofillers, which are used for both sealing and construction elements [2,5,6]. The main problem in obtaining polymer–wood composites is the lack of compatibility between the hydrophilic wood waste filler and hydrophobic polymer matrices. To ensure proper adhesion between the filler and matrix, it is important to treat the surface of wood waste to make it hydrophilic, for example by using acetic anhydride and thermal treatment [2,6–9]. Thermal treatment of wood waste causes changes in color, regrouping of polymers, and in the case of resinous trees, modification and redistribution of wood extracts [9,10]. The result is increased measurement stability and resistance to biodegradation, and lower mechanical properties. Previous research has shown that increasing the strength and stiffness is affected by the shape of the waste and not its size [11]. Wood–polymer composites may also prove to be alternatives to traditional materials used for facade panels, concrete fillers, packaging, and protective and anticorrosion coatings. These materials are especially interesting due to environmental concerns [1,2,6,12]. Silicone-based composites are one of the most important elastic technical materials produced industrially, e.g., glass-fiber-reinforced silicone composites [13]. In a high-power white light emitting diode (LED) package, the phosphor-silicone composite is typically used for photometric and colorimetric conversions, ultimately producing the white light [14]. Ceramifiable silicone rubber composites play important roles in the field of thermal protection systems (TPS) for rocket motor cases due to their advantages [15]. There are ceramizable (ceramifiable) silicone-based composites commonly used to increase flame retardancy of electrical cables and to ensure integrity of electricity network during fire by their ability to create a continuous ceramic structure [16]. In other paper silicone composites filled with different-sized nickel particles. The samples with particles showed larger improvements in shear storage modulus than those without particles [17]. By using an appropriate filler, composites with the desired properties can be obtained. The silicone-based membrane with 0.36 wt % of graphene oxide showed excellent antifouling performance, and is promising in practical applications [18]. There are no articles in the literature devoted to the introduction of fillers of natural origin into silicones. The authors took up this topic due to the need to create new materials based on waste products of natural origin.

This work aimed to assess the impact of wood waste from deciduous (beech, oak, and hornbeam) and coniferous trees (spruce) on the mechanical and biological properties of silicone composites used for structural element protection. Such a goal required the use of several physical and mechanical tests (static tensile test, hardness, abrasiveness, and density) and biological tests, which included cytotoxicity and aging tests in a replicated seawater environment.

2. Materials and Methods

2.1. Materials

The research was carried out on silicon matrix composites XIAMETER 4234-T4 (additive silicone) from Dow Corning (Table 1), in which the filler was wood waste from deciduous trees: oak (*Quercus robur*), beech (*Fagus sylvatica*), and hornbeam (*Carpinusbetulus*), and coniferous spruce (*Piceaabies*). Tree properties are summarized in Table 2. To identify the fillers, microscopic photos were taken on a stereo discovery Zeiss stereoscopic microscope (Carl Zeiss AG, Oberkochen, Germany), and single particles were imaged on a Zeiss Supra 35 scanning electron microscope, (Carl Zeiss AG, Oberkochen, Germany). The particle size distribution was tested on the Fritsch Analysette 22 Micro Tec Plus (FRITSCH GmbH, Idar-Oberstein, Germany).

Table 1. Properties of XIAMETER 4234-T4 silicone [19]. A—Xiameter RTV-4234-T-4 BASE; B—Xiameter T-4 curing agent.

Properties	Unit	Value
Ratio mixing A:B		10:1
Density	(g/cm^3)	1.1
Viscosity	(mPas)	35,000
Hardness	(ShA)	40
Linear shrinkage	(%)	<0.1
Tensile strength	(MPa)	6.7
Tensile strain	(%)	400

Table 2. Properties of wood fillers [20–23].

Properties	Oak	Beech	Spruce	Hornbeam
After drying density (kg/m^3)	900–1150	820–1270	700–850	660–1200
Flexural strength (MPa)	74–105	74–210	49–136	58–200
Flexural modulus (GPa)	10–13.5	10–18	7.3–21.4	7–17.7
Compressive strength (MPa)	48	41–99	30–79	54–99
Tensile strength (MPa)	50–180	55–180	21–245	24
Impact strength (J/cm^2)	1–16	3–19	1–11	8–12
Brinell's hardness (HBW)	34	34	12	29–36

2.2. Methodology

2.2.1. Composites

The composites were prepared by gravity casting with a content of 10 and 20 wt % wood waste. The mark of the composites is shown in Table 3.

Table 3. The mark of samples.

Name	Base	Filler	Filler Content (%)
XS		-	-
10 - O		oak	10
20 - O		oak	20
10 - B	XIAMETER 4234 - T4	beech	10
20 - B		beech	20
10 - S		spruce	10
20 - S		spruce	20
10 - H		hornbeam	10
20 - H		hornbeam	20

Before introducing wood waste into the silicone matrix, it was gradually heat-treated at 180°C for 180 min, until a constant weight was obtained. The dried waste was sieved to obtain particles smaller than 1 mm. Unification of the size of wood waste helped evenly distribute the filler in the obtained composites. Smaller waste sizes also promoted homogenous functional properties throughout produced composites. The silicone component A was mixed with the fillers on a high shear mixer, then the catalyst was added. The mixing speed was 500 RPM.

The prepared compositions were deaerated in a vacuum oven for 45 min at 0.78 bar and then cast into previously prepared, leveled molds to ensure a constant plate thickness of 5 mm. Seventy-two hours after being poured into molds, samples were cut by punching. The sample preparation scheme is shown in Figure 1.

The obtained samples were subjected to mechanical and biological tests. Mechanisms were examined using hydrostatic weighing tests, radiation resistance tests, Schopper abrasion resistance tests, and Shore type A hardness tests and static tensile tests. Toxicity to normal human cells was also evaluated using the MTT test. The possibility of using the received WPC to obtain cable covers, which are placed on the seabed was also examined. All tests were carried out at the temperature in the room where the test was performed, which was 22 °C with a humidity of 50%.

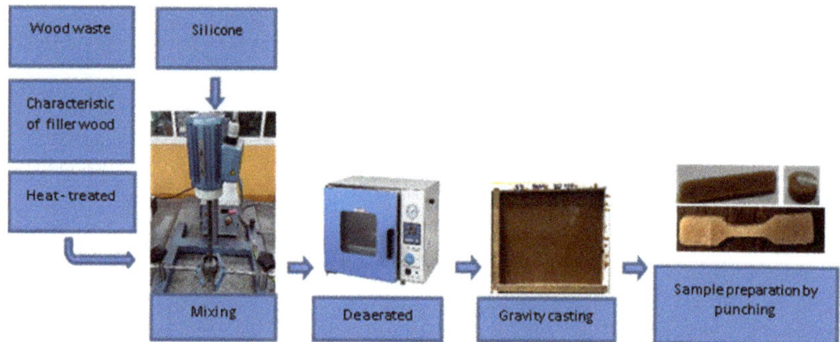

Figure 1. Scheme of preparation samples.

2.2.2. Density Testing by Hydrostatic Weighing

Densities were measured on a scale in accordance with EN ISO 1183-1:2006 using 5 samples from each system [24]. The method of determining the density by the method of hydrostatic weighing of composite polymer materials consisted in weighing the test sample with the use of an OHAUS Adventurer-Pro (OHAUS Europe GmbH, Nänikon, Switzerland) analytical balance with a density measurement kit. The sample was weighed twice. The first measurement is carried out with the sample placed on the pan and surrounded by air. The second measurement is carried out for a sample immersed in a liquid of known density. During the tests, water was used as a liquid with a known density $d = 0.997$ g/cm^3.

Density was determined using Formula (1):

$$d = d_{H20} \frac{m_1}{(m_1 - m_2)} \tag{1}$$

where:

- d_{H20}—density of water (g/cm^3),
- m_1—dry sample mass (g),
- m_2—wet sample mass (g).

2.2.3. Rebound Resilience with Schober's Test

The resilience test was carried out in accordance with the EN ISO 4662:2017 standard for five samples with dimensions of 30 mm × 30 mm × 5 mm [25]. Before conducting basic tests, the samples were mechanically conditioned (2 impacts). The measurement of the resilience of composite materials consists in hitting the sample with a weight placed on a pendulum. The sample is held in an anvil attached to a metal body. The measurement consisted in reading the value indicated by the pointer on the value axis (%).

2.2.4. Abrasion Resistance Tests

The abrasion resistance tests were carried out on a Schopper–Schlobach apparatus (APGI) in accordance with the ISO 4649:2007 standard [26]. In the research, sandpaper (60 grit) was used, wound on a roller with a diameter of 150 mm, which was rotating at a speed of 40 RPM. Abrasion resistance was determined for 3 cylindrical samples with a 16 mm diameter and 10 mm height. Due to the thickness of spilled boards (5 mm), the samples were glued with cyanoacrylate adhesive (according to ISO 4649: 2007). Abrasion resistance (abrasive wear), i.e., the volume loss relative to a standard sample, was determined based on the Formula (2):

$$\Delta V = \frac{m_1 - m_2}{d} \tag{2}$$

where:

- m_1—mass of sample before abrasion (g),
- m_2—sample mass after abrasion (g),
- d—sample density (g/cm^3).

2.2.5. Hardness Test

The shore A hardness test was carried out in accordance with ISO 7619-1:2010 [27]. The measurements were made with a hardness durometer Shore A type (Etopoo). Five measurements were taken on each of the composites, maintaining a distance of at least 10 mm from the sample edge.

2.2.6. Tensile Test

Tensile strength tests were performed in accordance with EN ISO 527-1 [28]. The measurements were carried out according to EN ISO 527-1 [28] for 5 samples (type 5-B) cut from each composition and native samples. The test was carried out on the Instron 4465 tensile test machine. The test speed was 50 mm/min. The stress at break and strain at break were determined.

2.2.7. Cytotoxicity Testing of Composite Samples

Cytotoxicity testing of composite samples according to the procedure described earlier [29,30]. Cell viability of normal human dermal fibroblasts (NHDF, Lonza) was assessed using an MTT (3-[4-5-dimethylthiazol-2-yl]-2,5-diphenyltetrazolium bromide) test. Cells were seeded in Petri plates with a concentration of 10^5 cells per well. Cell cultures were seeded on tested materials and incubated for 72 h at 37 °C in a humidified atmosphere with 5% CO$_2$. Then, the culture medium was removed and replaced with a trypsin solution for cell collection. After trypsin neutralization, cell suspensions were centrifuged (2000 RPM, 3 min, room temperature) and cell pellets were resuspended in the MTT solution (50 μL, 0.5 mg/mL in RPMI 1640 without phenol red, Sigma, (Saint Louis, MO, USA). After 3 h of incubation, the MTT solution was removed, and the acquired formazan was dissolved in isopropanol:HCl system. Finally, the absorbance at 570 nm was spectrophotometrically measured with a plate reader Epoch, BioTec, (Winooski, VT, USA). The results were expressed as a survival fraction (%) in comparison to the untreated controls (100%), from 3 separate experiments, ±SD.

2.2.8. Ageing in Seawater Conditions

Due to the potential application of the prepared materials for the covers of structures installed in the seabed, tests were carried out in the sea water environment. The possibility of surface settling of obtained composites was investigated using marine algae. The tests were carried out in an aquarium with samples mounted on a wooden plate. Seawater was prepared in accordance with the standard ASTM D 1141-52 [31]. The chemical composition of the substitute seawater is shown in Table 4.

Table 4. Chemical composition of substitute sea water [31].

Ingredients	Concentration (g/L)
Sodium chloride (NaCl)	24.53
Magnesium chloride (MgCl$_2$)	5.2
Sodium sulfate (Na$_2$SO$_4$)	4
Calcium chloride (CaCl$_2$)	1.16
Potassium chloride (KCl)	0.695
Sodium bicarbonate (NaHCO$_3$)	0.201
Potassiumbromide (KBr)	0.101
Boric acid (H$_3$BO$_3$)	0.027
Strontium chloride (SrCl$_2$)	0.025
Sodium fluoride (NaF)	0.003

Marine algae was added to the aquarium, which was intensively illuminated for 30 days. After this time, the sample plate was removed. The aging was carried out at the temperature of 30 °C, while illuminating the aquarium with a reflector placed 1 m from the aquarium.

3. Results and Discussion

3.1. Characteristics of Fillers

Wood waste fillers are shown in Figure 2. Images of wood fillers were taken on a stereo discovery Zeiss stereoscopic microscope (Figure 2a,c,e,g) and single filler particles were imaged on a Zeiss Supra 35 scanning electron microscope (Figure 2b,d,f,h). Particle size tests were performed on a Fritsch Analysette 22 Micro Tec Plus equipped with a wet dispersing unit. Measuring range 0.08–2000 microns. The particle size distribution is shown in Figure 3. The frequency curves were developed on the basis of 5 measurements for each wood waste. As can be seen, the smallest filler particles were observed for beech and hornbeam.

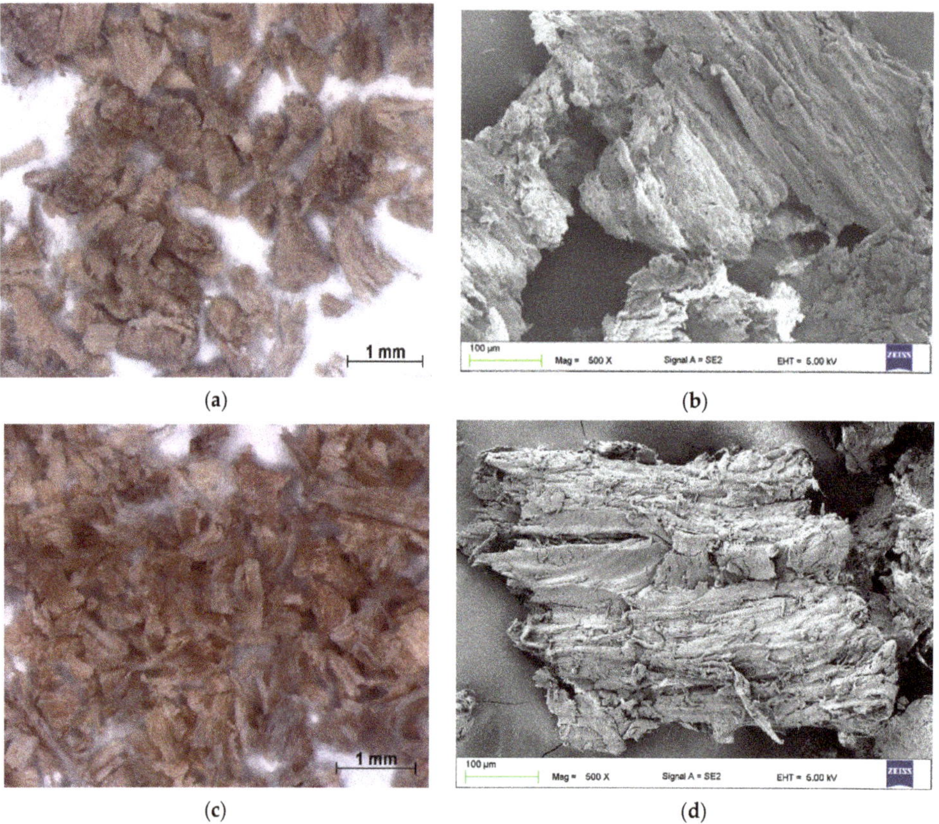

Figure 2. *Cont.*

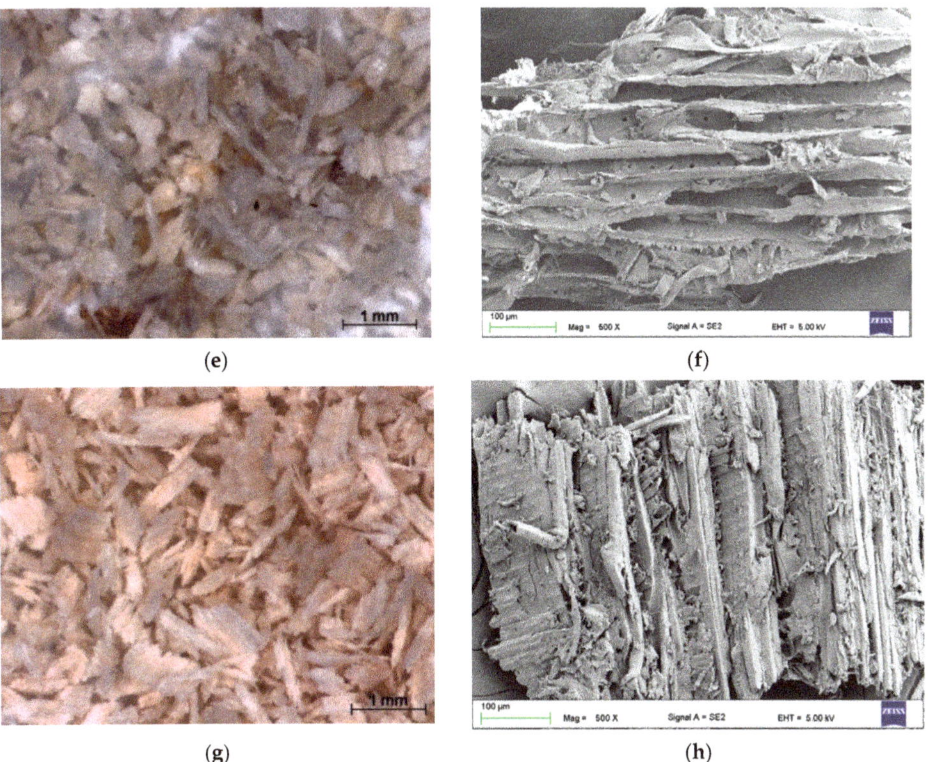

Figure 2. Microscopic image oak (**a**,**b**), beech (**c**,**d**), spruce (**e**,**f**), and hornbeam (**g**,**h**).

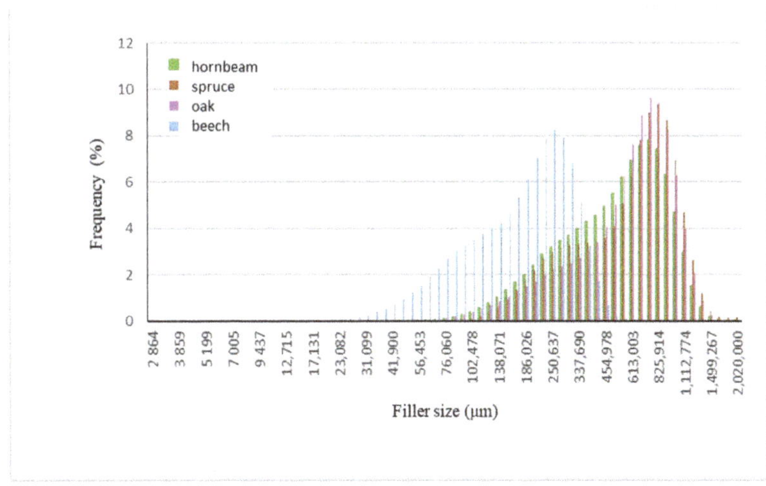

Figure 3. The frequency curve for the tested fillers.

Analyzing the images of fillers presented in Figure 2, it can be noticed that the structure of spruce differs from that of other trees. Spruce has a tubular structure. The structure of spruce is characterized by visible tracheas and pits, which, for example, increases the mechanical adhesion between the matrix and the filler. However, the release of resin without proper preparation of the spruce filler may adversely affect some properties. Microscopy images of fillers used in this study showed no significant differences in shape, but differed in surface image. These differences can primarily be seen between deciduous tree waste and spruce.

The size of the most common particles was 747.78 µm for beech (8.2%), hornbeam (7.8%), and oak (9.6%), and 825.9 µm for spruce (9.3%). As can be seen, the smallest particles are characteristic of beech (90%—940.03 µm) and hornbeam (90%—917.03 µm), while the largest range was noted for oak (90%—998.13 µm), while for spruce 90% of the particles have the size below 971.77 µm (Figure 4). This is due to the structure and properties of individual trees.

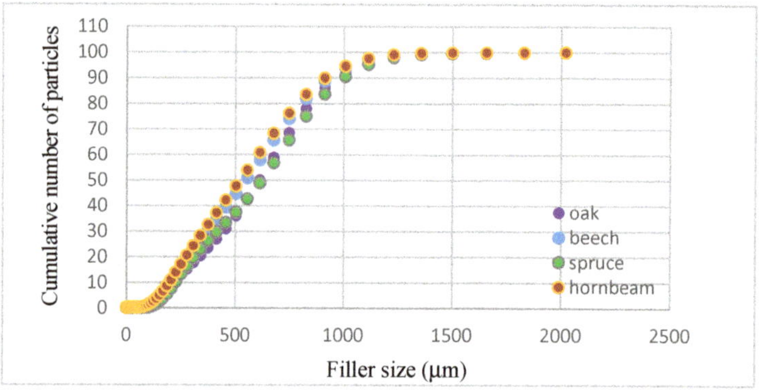

Figure 4. Cumulated curves of the filler's size.

3.2. Density Test Results

The density of composites produced using wood waste filler were determined by measuring five samples. The average values and their standard deviations are graphically displayed in Figure 5.

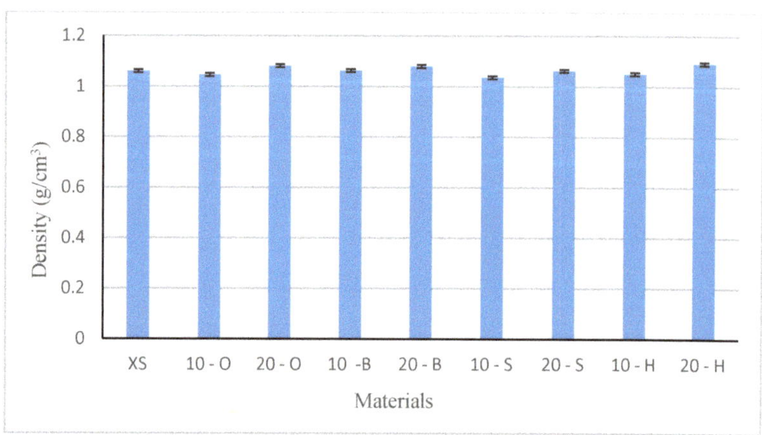

Figure 5. Densities of tested materials (g/cm^3).

Samples from XIAMETER 4234-T4 silicone showed a density of 1.069 g/cm^3; the specified density was 1.1 g/cm^3. The introduction of 10% filler, regardless of the type of wood waste, reduced the density of the obtained composites, and the lowest density was obtained using spruce wood waste. The density of the spruce-silicone composite decreased by approximately 5% compared to the native sample (XS). It was most likely caused by the structure of spruce waste as shown in Figure 2f. This is probably the result of tracheas (Figure 2f) distributed throughout the wood, which is why spruce is considered to be a light tree (range 410–500 kg/m^3, and after drying 700–850 kg/m^3) and the measurement method. The highest density at 10% filling was recorded for hornbeam, approximately 3% in relation to spruce. This is mainly due to the density and structure of the wood. The hornbeam has a uniform and fine structure, which affects its properties. Hornbeam, beech, and oak belong to heavy and very heavy trees, whose densities after reach 1200 kg/m^3 (Table 2). Buk contains numerous wood rays, which serve as conductive channels. Oak wood, unlike beech wood, is characterized by wide conductive channels but in much smaller quantities. Composites in which the share of wood waste was 20% had a similar or slightly higher density than native silicone. Composites with 20% filling with hornbeam waste were characterized by the highest density by approximately 4% compared to XS. The conducted one-way analysis of variance showed a significant influence of the examined factors on the density of composites (F = 15.71, test F = 2.208).

3.3. Rebound Resilience Results

The average resilience values and standard deviations of the tested materials are shown in Figure 6.

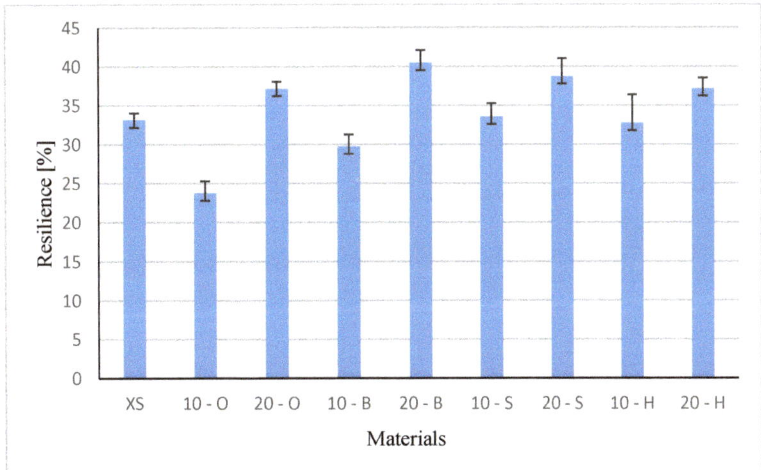

Figure 6. Resilience values of tested materials (%).

Introduction 10 wt % wood waste caused a reduction in elasticity for composites with oak by 27% and beech by 20% in relation to native samples, but no change was observed in composites with spruce filling, while a slight increase was observed in the resilience of samples with hornbeam filling. Such a reduction may be related to the size of the grain introduced into the composite and the structure of the wood. The beech particles were characterized by the smallest dimensions, which are confirmed by the normal distribution—Figure 3. Although the literature reports [11] a greater impact on the properties of composites, the shape of the particle has a greater impact, and in the examined case the size is also important. It seems that this is also the result of obtaining the filler from the waste and the properties of the tree itself.

In all types of composites containing 20 wt % filling the wood, resilience increased, the highest of which was observed in samples with beech filling (41%). In composites with hornbeam filling, the resilience results were relatively similar (34.6 and 36.6%). Hornbeam wood is very hard and difficult to break, hence the resilience remained at a similar level, regardless of the content. Regardless of the wood's cleavage and its hardness, the additional filler improved the flexibility in all systems containing 20 wt % filler. It can therefore be concluded that the introduction of 20 wt % of the wood filler into the systems, regardless of the wood type, its structure, and the particle size, improved the resilience of the obtained composite. That is why the samples recovered their original shape after applying more force than in the unmodified system. The carried out one-way analysis of variance showed that both the content and the type of filler had a significant impact on the change of relativity (F = 38.47, F = 2.208 test).

3.4. Abrasion Resistance Results

The results of abrasion resistance tests were compared to a standard sample made of silicone XIAMETER 4234-T4 without filler. The average abrasion resistance values are shown in Figure 7. The test results show that the introduction of filler into the matrix reduced abrasion. The composite containing 10 wt % beech waste (10 - O) showed a loss of 0.51 cm^3 and had the highest loss of volume. In the case of fillers obtained from deciduous trees, abrasion is approximately 20% lower than that of the native sample. Spruce waste causes a loss of volume by 65% compared to the native sample. The introduction of 20% beech and hornbeam fillers did not affect the abrasion resistance. The differences observed fall within the potions of errors. However, it can be assumed that the distribution of filler particles in the structure of the composite resulted in no changes observed. These fillers are characterized by the smallest particles: beech 90%—940.03 μm and hornbeam 90%—917.03 μm (Figure 4). The lowest abrasion (approximately 50% of XS abrasion) was observed for spruce waste, similarly to composite materials with 10% filling. It seems that the related abrasion is determined mainly by the type of wood and particle size. The smaller the particle, the greater the loss in volume. Samples filled with deciduous tree waste are characterized by worse abrasion resistance than composites with spruce waste. The conducted one-way analysis of variance confirms the strong relevance of the studied factors for the abrasion of composites (F = 202.9, test F = 2.51).

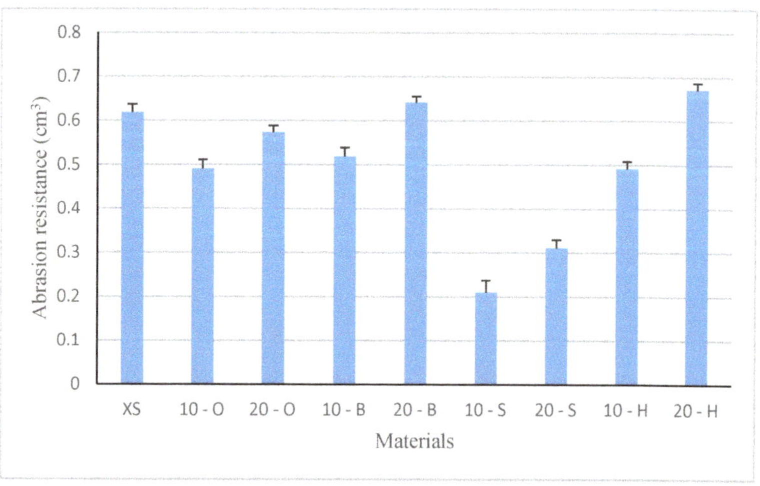

Figure 7. Abrasion resistance of composites (cm^3).

3.5. Shore A Hardness Test

Five samples of each composite variant were tested for hardness tests. Graphical presentations of hardness test results and standard error are presented in Figure 8.

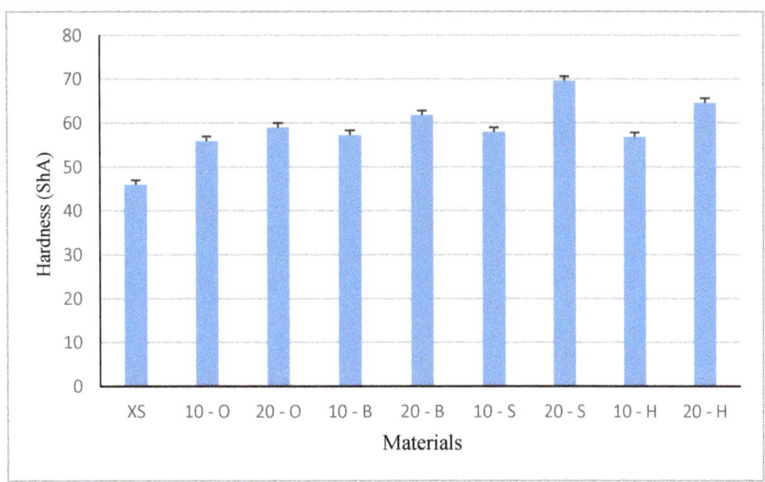

Figure 8. Hardness value of samples (ShA).

In each modified system, the obtained hardness was higher than the unmodified sample. In systems containing 10 wt % XIAMETER 4234-T4 silicone, the obtained composites had comparable hardness values. Both heavy woods like beech, hornbeam, oak, and light spruce showed a hardness of around 57 ShA. Such a high value in the case of spruce precipitation may indicate the content of the resin inside the waste. As in previous systems, when 20 wt % filler was used, the highest value was obtained in samples containing spruce waste. The obtained value of 69.6 ShA was 11 ShA higher than hard oak waste samples.

This may prove that the separated spruce wood resin additionally increased the hardness of the composite, which was also confirmed by the results of resilience. Relatively soft spruce wood did not differ from harder beech, hump, or oak, even taking into account the particle size. As can be seen, as in previous studies, both the type of wood waste and the size of the particle had a significant impact on the obtained hardness results (F = 91.1, test F = 2.208).

3.6. Tensile Test Results

Tensile test samples were cut using a punch, which ensured a repeatable shape and dimensions of tested samples. Five measurements were made for each composite variant (filler fraction, type of wood waste). A graphical presentation of the results is shown in Figure 9 (strain at break) and Figure 10 (stress at break).

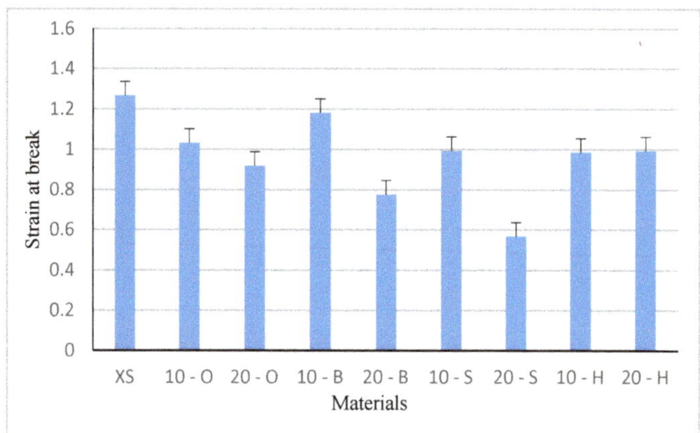

Figure 9. Strain at break.

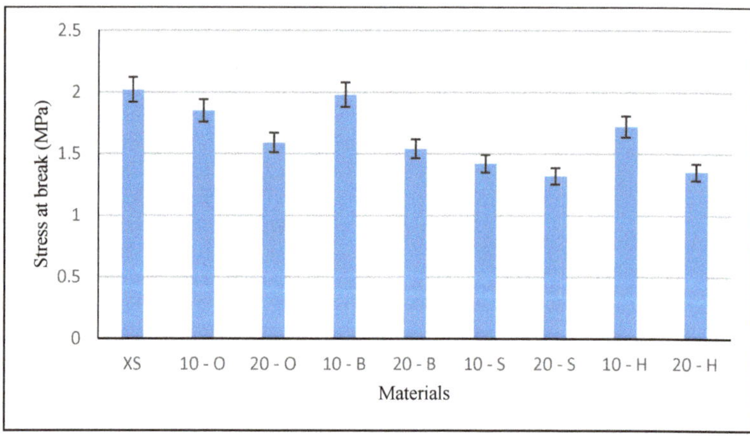

Figure 10. Stress at break (MPa).

Samples of pure XIAMETER 4234-T4 silicone showed the greatest strain at break, and all other composites showed reduced strain at break. The sample showing the most similar strain to unmodified silicone (1.269) was that with 10 wt % beech content (1.181). The lowest value was observed in composite 20 - S (0.567), but also the highest measurement error of 6% and standard deviation of 0.06, which may be a result of the preparation of the filler for introduction, the introduction technology itself or the physicochemical properties of wood. Composites containing hornbeam filling showed similar deformation values, irrespective of the percentage of the composite (0.994 and 0.985, respectively). This seems to be related primarily to the distribution of the filler in the composite, and secondly to the grain size. The one-way analysis of variance showed a significant impact on the stain values of the type and content of the filler (F = 162.19, test F = 2.208).

It can be observed that the smaller the filler particles and the proportion by weight, the better the strain at break, which confirms the adopted theories. The lowest strain at break was observed for composites with 20% spruce filling, which proves a significant influence of wood properties on the tested composite characteristics.

The tensile test results indicate that a higher stress at break was observed in samples modified with XIAMETER 4234-T4 silicone. All composites showed lower stress at break, with the exception of composite 10 - B, whose value (1.98 MPa) was similar to unmodified silicone (2.02 MPa). The greatest differences between the native and filled samples were observed for composites with meringue waste and amounted to 10 wt %—30% and for 20 wt %—35%. The composites with the filler from deciduous trees had smaller differences. For 10%, the filling was on average approximately 8% (the lowest value was shown by hornbeam—1.72 MPa), while for 20% by wt filling 26%. ANOVA analysis confirmed the significant influence of the tested factors on the stress at break, F = 35.75, test F = 2.208.

3.7. Cytotoxicity Test Results

The cytotoxicity of the tested composites was obtained according to the standard procedures against model cell lines [29,30,32]. The viability and proliferation assays, such as an MTT assay, allowed for biocompatibility of different, natural, or synthetic agents assessments [32,33]. Figure 11 presents the cytotoxicity against NHDF cells. The cytotoxicity test results showed that neither silicon nor any of the tested composites showed any toxicity to the normal fibroblasts. After 72 h, the fraction of living cells for the XS, 10 - B, and 10 - O groups showed similar values of 100%, 98%, and 93%, respectively. The remaining results showed significant cell proliferation when in contact with composites. The 20 - O and 20 - H composites showed the most proliferative values, for which the living cell fractions were 190% and 196%, respectively. The results confirm the biocompatibility of composites and any toxicity against the NHDF cell line in a standard in vitro cytotoxicity assay [34] was reported.

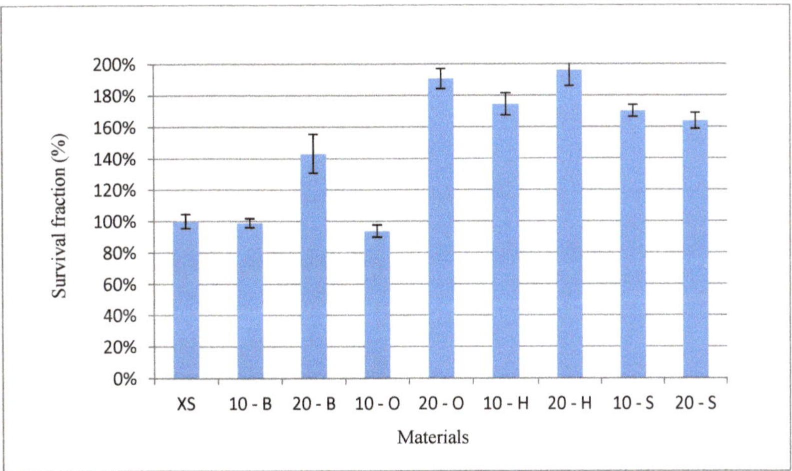

Figure 11. Cytotoxicity test results (%).

3.8. Possibly of Using WPC on the Seabed

Figure 12 shows the image of composite samples before and after ageing in seawater.

Based on the observation of plates covered with marine algae, it can be concluded that algae covered the composites with a higher wood waste content. The exception is the composite containing hornbeam waste in which higher algae content was observed on the sample containing 10 wt % wood waste. The silicone sample was also overgrown by marine algae. Most importantly, in all the samples tested, there is a larger or smaller sample of marine algae. After performing additional tests, this will enable the use of the tested composites as materials for use on the seabed. The results of this research are presented in Table 5.

Table 5. Results of microscopic research: wood filler and aged samples.

Wood Filler	10%	20%
oak	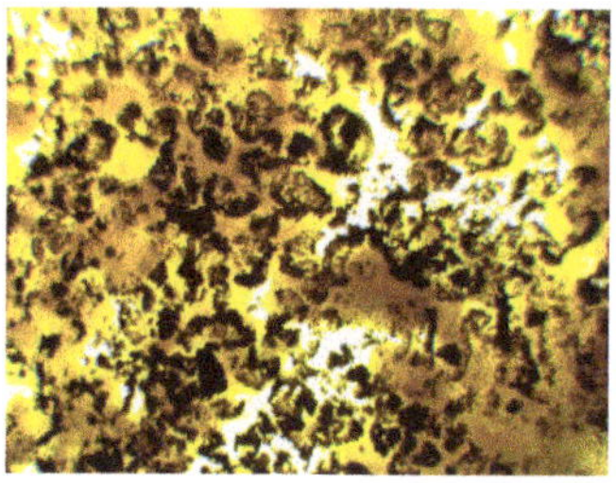	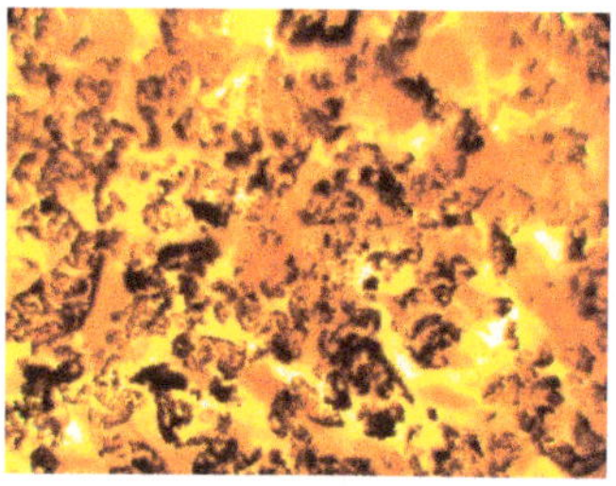
beech	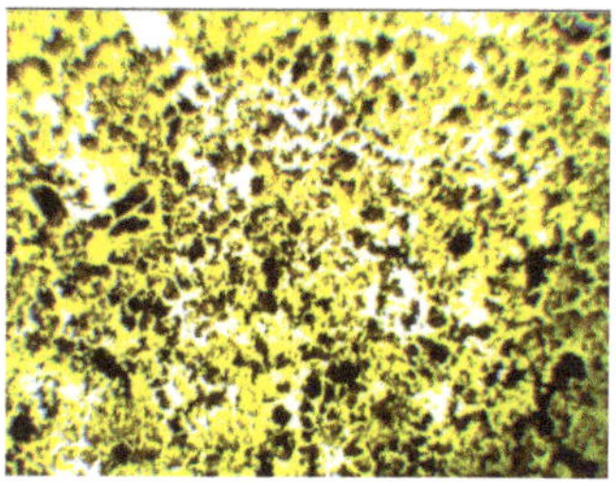	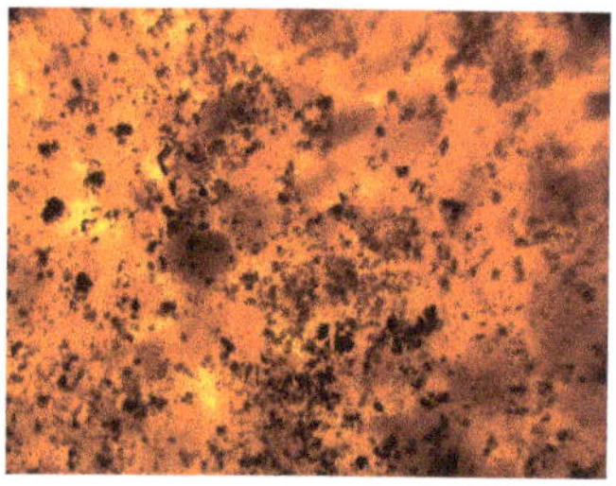
spruce	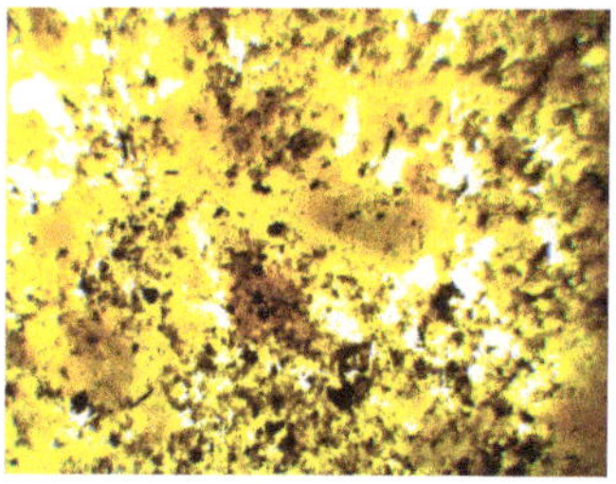	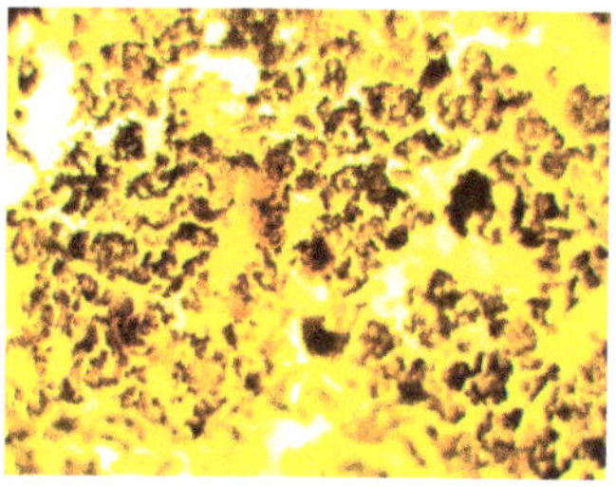
hornbeam	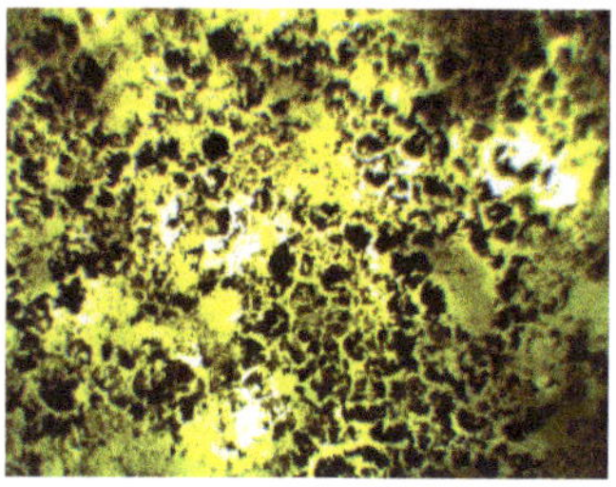	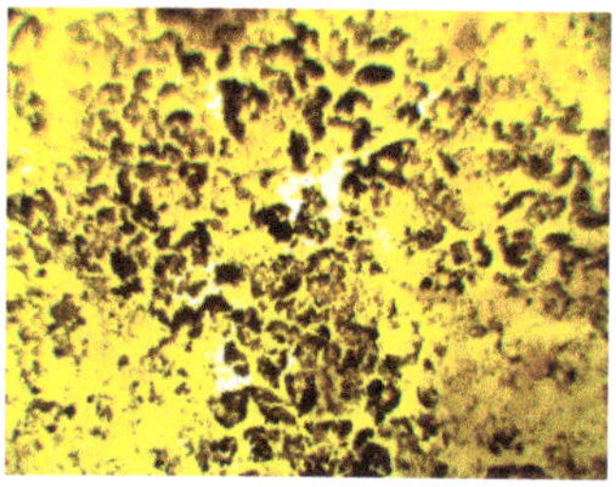

Microscopy images of samples aged in seawater showed the growth of marine algae film on their surface. Along with filler content, the number of marine algae increased and the color of the silicone matrix changed. At the same time, it should be noted that in the case of spruce waste, a smaller algae growth rate was observed compared with composites filled with deciduous tree waste. Samples with 10 wt % filler showed more diffuse growth than those with 20 wt % filler. Composites containing hornbeam waste were the most overgrown, while the least overgrown were those with beech filling, possibly due to the hardness of the tree (beech shows the lowest hardness among deciduous trees). This also explains the observations of the spruce composite filling. It can therefore be concluded that more algae growth occurred on harder trees (i.e., those containing more cellulose fibers).

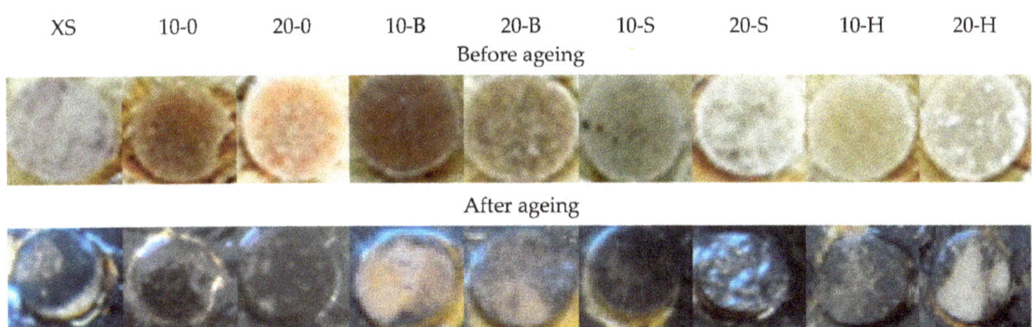

Figure 12. Comparison of plates with composites before and after immersion in a marine aquarium.

In order to select the best filler and content, a simplified multicriteria assessment table was developed (Table 6). The grading scale was adopted from 1 to 5, where 5 means the composites with the best properties and 1 the worst. The same grade was assigned for identical averaged test results. Assessments were made for each percentage of filling separately, based on the results of the tests carried out. The results of the cytotoxicity and ageing tests were not included in the multicriteria analysis. Cytotoxicity measured in the MTT test showed no cytotoxicity of all materials against normal cells, NHDF and more accurate interaction of materials with cells requires additional tests. In turn, the aging was assessed on the basis of macro and microscopic photos of the surface, which makes it impossible at this stage to assess changes in the mechanical and physicochemical properties of the materials.

Table 6. Quality assessment of wood–silicon composites, W_c—criterion weight, E—evaluation, R—result.

Własność	W_c	XS E	XS R	10-O E	10-O R	20-O E	20-O R	10-B E	10-B R	20-B E	20-B R	10-S E	10-S R	20-S E	20-S R	10-H E	10-H R	20-H E	20-H R
Density	1	5/3	5/3	3	3	4	4	5	5	4	4	2	2	3	3	4	4	5	5
Resilience	4	1	4	2	8	3	12	5	3	5	20	5	20	4	16	4	16	3	12
Abrasion resistance results	6	1	6	4	24	3	18	3	18	2	12	5	30	5	30	4	24	2	12
Hardness	3	1	3	2	6	2	6	4	12	3	9	5	15	5	15	3	9	4	12
Strain at break	2	5	10	3	6	3	6	4	8	2	4	2	4	2	4	1	2	4	8
Stress at break	5	5	25	3	15	4	20	4	20	3	15	1	5	1	5	2	10	2	10
Suma			53/51		62		66		66		64		76		73		65		59

4. Conclusions

The conducted research allows the following conclusions to be reached:

The density of composites changed with filler content and the type of wood. A reduction of approximately 3–5% in density was observed for the 10% fill composites and for 20% fill and an increase of up to 4%. This is mainly due to the structure of wood waste and the density of fillers. However, on the basis of statistical analysis, it was found that the examined variables are of little importance for the tested characteristic.

The introduction of wood filler into composites increased the hardness, which is not directly proportional to the hardness of trees. The highest hardness was characteristic for the composite with 10 and 20% spruce filler (density 700–850 kg/m^3), which may be related to the particle size of 998.13 µm and the resin presence probably additionally hardening the composite. At the same time, the silicone-spruce composite shows the smallest abrasive wear by an average of approximately 30% compared to deciduous trees fillers. Both the size and type of the tree from which the waste was obtained significantly influences the hardness and abrasion.

For 20% at the higher the resilience and the lower the strain and stress at break was observed. All composites with 20 wt % filling, showed higher resilience and lower values of the characteristics determined in the static tensile test. For the 10% share, there is no unequivocal dependence of the resilience on the filler content in the composite. It seems that in this case both the type of wood and the property gradient resulting from the casting technology are important. The tested characteristics are strongly dependent on the weight fraction of the filler and the properties of the wood waste.

Conducted aging tests in the sea water environment showed that the area covered with algae increased with the increase of the filler, which was related to the type and structure of wood waste. All composites stimulated the proliferation of normal cells of the human body, demonstrating their lack of toxicity on normal fibroblasts, NHDF cells.

The multicriteria analysis proves that the best test results were obtained for composites filled with spruce waste, both with 10% by weight and with 20% by weight.

Further research will allow for a more complete characterization of WPC as materials for use in protective coatings, insulation systems, or packaging.

Author Contributions: M.M. and M.S. (Małgorzata Szymiczek) conceived, designed and carried out the experiments, analyzed data and wrote the paper. M.S. (Magdalena Skonieczna) was responsible for cytotoxicity analysis. All authors were involved in revising the paper's important content. All authors have read and agreed to the published version of the manuscript.

Funding: This research received no external funding.

Acknowledgments: The authors would like to thank Mateusz Lipiarz (a student doing the master's thesis) for his help in carrying out the research. Mikołaj Mrówka (AGH University of Science Technology, Poland) is acknowledged for help with Figures formatting.

Conflicts of Interest: The authors declare no conflict of interest.

References

1. Ashori, A. Wood–plastic composites as promising green-composites for automotive industries! *Bioresour. Technol.* **2008**, *99*, 4661–4667. [CrossRef] [PubMed]
2. Klyosov, A.A. *Wood—Plastic Composites*, 1st ed.; John Wiley & Sons: Hoboken, NJ, USA, 2007; pp. 15–37.
3. Clemons, C. Wood-plastic composites in the United States: The interfacing of two industries. *Forest Prod. J.* **2002**, *52*, 10–18.
4. Pritchard, G. Two technologies merge: Wood plastic composites. *Plast. Addit. Compd.* **2004**, *6*, 18–21. [CrossRef]
5. Cui, Y.; Lee, S.; Noruziaan, B.; Cheung, M.; Tao, J. Fabrication and interfacial modification of wood/recycled plastic composite materials. *Compos. Part A Appl. Sci. Manuf.* **2008**, *39*, 655–661. [CrossRef]
6. Najafi, S.K. Use of recycled plastics in wood plastic composites—A review. *Waste Manag.* **2013**, *33*, 1898–1905. [CrossRef]
7. Rowell, R.M. Acetylation of wood: A journey from analytical technique to commercial reality. *For. Prod. J.* **2006**, *56*, 4–12.
8. Rowell, R.M.; Ibach, R.E.; McSweeny, J.; Nilsson, T. Understanding decay resistance, dimensional stability and strength changes in heat-treated and acetylated wood. *Wood Mater. Sci. Eng.* **2009**, *1–2*, 14–22. [CrossRef]
9. Lahtela, V.; Kärki, T. Effects of impregnation and heat treatment on the physical and mechanical properties of Scots pine (*Pinus sylvestris*) wood. *Wood Mater. Sci. Eng.* **2014**, *11*, 217–227. [CrossRef]
10. Nuopponen, M.; Vuorien, T.; Jamsa, S.; Viitaniemi, P. The effects of a heat treatment on the behaviour of extractives in softwood studied by FTIR spectroscopic methods. *Wood Sci. Technol.* **2003**, *37*, 109–115. [CrossRef]
11. Stark, N.M.; Rowlands, R.E. Effect of wood fiber characteristics on mechanical properties of wood/polypropylene composites. *Wood Fiber Sci.* **2003**, *35*, 167–174.
12. Winandy, J.E.; Stark, N.M.; Clemons, C.M. Considerations in recycling of wood-plastic composites. In Proceedings of the 5th Global Wood and Natural Fibre Composites Symposium, Kassel, Germany, 27–28 April 2004.
13. Beter, J.; Schrittesser, B.; Lechner, B.; Reza Mansouri, M.; Marano, C.; Fuchs, P.F.; Pinter, G.; Beter, J.; Schrittesser, B.; Lechner, B.; et al. Viscoelastic Behavior of Glass-Fiber-Reinforced Silicone Composites Exposed to Cyclic Loading. *Polymers* **2020**, *12*, 1862. [CrossRef] [PubMed]
14. Fan, J.; Wang, Z.; Zhang, X.; Deng, Z.; Fan, X.; Zhang, G. High Moisture Accelerated Mechanical Behavior Degradation of Phosphor/Silicone Composites Used in White Light-Emitting Diodes. *Polymers* **2019**, *11*, 1277. [CrossRef] [PubMed]

15. Song, J.; Huang, Z.; Qin, Y.; Wang, H.; Shi, M. Effects of Zirconium Silicide on the Vulcanization, Mechanical and Ablation Resistance Properties of Ceramifiable Silicone Rubber Composites. *Polymers* **2020**, *12*, 496. [CrossRef] [PubMed]
16. Imiela, M.; Anyszka, R.; Bieliński, D.M.; Pędzich, Z.; Zarzecka-Napierała, M.; Szumera, M. Effect of carbon fibers on thermal properties and mechanical strength of ceramizable composites based on silicone rubber. *J. Therm. Anal. Calorim.* **2016**, *124*, 197–203. [CrossRef]
17. Song, P.; Peng, Z.-J.; Yue, Y.-L.; Zhang, H.; Zhang, Z.; Fan, Y.-C. Mechanical properties of silicone composites reinforced withmicron- and nano-sized magnetic particles. *eXPRESS Polym. Lett.* **2013**, *7*, 546–553. [CrossRef]
18. Jin, H.; Bing, W.; Tian, L.; Wang, P.; Zhao, J. Combined Effects of Color and Elastic Modulus on Antifouling Performance: A Study of Graphene Oxide/Silicone Rubber Composite Membranes. *Materials* **2019**, *12*, 2608. [CrossRef]
19. XIAMETER®RTV-4234-T4 Base and XIAMETER®T4/T4 OCuring Agent. Available online: https://www.notcutt.co.uk/wp-content/uploads/2014/06/rtv-4234-t4-base-t4-t4-0-cat-sales-lit.pdf (accessed on 10 December 2020).
20. Bruechert, F.; Becker, G.; Speek, T. The mechanics of Norway spruce [Piceaabies]: Mechanical properties of standing trees from different thinning regimes. *For. Ecol. Manag.* **2000**, *35*, 45–62. [CrossRef]
21. Kelley, S.S.; Rials, T.G.; Snell, R.; Groom, L.H.; Sluiter, A. Use of near infrared spectroscopy to measure the chemical and mechanical properties of solid wood. *Wood Sci. Technol.* **2004**, *38*, 257–276. [CrossRef]
22. Plomion, C.; Leprovost, G.; Stokes, A. Wood Formation in Trees. *Plant Physiol.* **2001**, *127*, 1513–1527. [CrossRef]
23. Wessels, C.B.; Malan, F.S.; Rypstra, T. A review of measurement methods used on standing trees for the prediction of some mechanical properties of timber. *Eur. J. For. Res.* **2011**, *130*, 881–893. [CrossRef]
24. EN ISO. *1183-1: 2006 Plastics—Methods for Determining the Density of Non-Cellular Plastics—Part 1: Immersion Method, Liquid Pyknometer Method and Titration Method*; International Organization of Standardization: Geneva, Switzerland, 2006.
25. ISO. *4662:2017 Rubber, Vulcanized or Thermoplastic—Determination of Rebound Resilience*; Springer: Cham, Switzerland, 2017.
26. EN ISO. *4649:2007 Rubber, Vulcanized or Thermoplastic—Determination of Abrasion Resistance Using a Rotating Cylindrical Drum Device*; International Organization of Standardization: Geneva, Switzerland, 2007.
27. EN ISO. *7619-1:2010 Rubber, Vulcanized or Thermoplastic—Determination of Indentation Hardness—Part 1: Durometer Method (Shore Hardness)*; International Organization of Standardization: Geneva, Switzerland, 2010.
28. EN ISO. *527-1:2012 Plastics—Determination of Tensile Properties—Part 1: General Principles*; International Organization of Standardization: Geneva, Switzerland, 2012.
29. Mrówka, M.; Jaszcz, K.; Skonieczna, M. Anticancer activity of functional polysuccinates with N-acetyl-cysteine in side chains. *Eur. J. Pharmacol.* **2020**, *885*, 173501. [CrossRef] [PubMed]
30. Mrówka, M.; Szymiczek, M.; Machoczek, T.; Lenża, J.; Matusik, J.; Sakiewicz, P.; Skonieczna, M. The influence of halloysite on the physicochemical, mechanical and biological properties of polyurethane-based nanocomposites. *Polimery* **2020**, *65*, 784–791. [CrossRef]
31. American Society for Testing and Materials. *Standard Practice for the Preparation of Substitute Ocean Water*; ASTM International: West Conshohocken, PA, USA, 2013.
32. Skonieczna, M.; Hudy, D.; Hejmo, T.; Buldak, R.J.; Adamiec, M.; Kukla, M. The adipokine vaspin reduces apoptosis in human hepatocellular carcinoma (Hep-3B) cells, associated with lower levels of NO and superoxide anion. *BMC Pharmacol. Toxicol.* **2019**, *20*, 58. [CrossRef] [PubMed]
33. Adamiec, M.; Skonieczna, M. UV radiation in HCT 116 cells influences intracellular H_2O_2 and glutathione levels, antioxidant expression, and protein glutathionylation. *Acta Biochim. Pol.* **2019**, *66*, 605–610. [CrossRef]
34. Altmann, S.; Choroba, K.; Skonieczna, M.; Zygadło, D.; Raczyńska-Szajgin, M.; Maroń, A.; Małecki, J.G.; Szłapa-Kula, A.; Tomczyk, M.; Ratuszna, A.; et al. Platinum(II) coordination compounds with 4′-pyridyl functionalized 2,2′:6′,2″-terpyridines as an alternative to enhanced chemotherapy efficacy and reduced side-effects. *J. Inorg. Biochem.* **2019**, *201*, 110809. [CrossRef]

MDPI

St. Alban-Anlage 66

4052 Basel

Switzerland

Tel. +41 61 683 77 34

Fax +41 61 302 89 18

www.mdpi.com

Polymers Editorial Office

E-mail: polymers@mdpi.com

www.mdpi.com/journal/polymers